WAVELENGTH STANDARDS IN THE INFRARED

WAVELENGTH STANDARDS
IN THE INFRARED

K. NARAHARI RAO

Department of Physics
The Ohio State University
Columbus, Ohio

CURTIS J. HUMPHREYS

Research Department
U. S. Naval Ordnance Laboratory
Corona, California

D. H. RANK

Department of Physics
The Pennsylvania State University
University Park, Pennsylvania

1966

ACADEMIC PRESS · NEW YORK · LONDON

ACADEMIC PRESS INC.
111 Fifth Avenue, New York, New York 10003

United Kingdom Edition published by
ACADEMIC PRESS INC. (LONDON) LTD.
Berkeley Square House, London W. 1

LIBRARY OF CONGRESS CATALOG CARD NUMBER:65-26044

PRINTED IN THE UNITED STATES OF AMERICA

PREFACE

The importance of the infrared region of the electromagnetic spectrum to research in chemistry, physics, and engineering is well known. During the past thirty years, study of the infrared spectra of molecules has yielded information on the structures of molecules such as, the geometrical arrangements of the nuclei in molecules and precise values for internuclear distances. In the past decade, many significant advances have been made in the theory of polyatomic molecular spectra which, given good basic data, permit the effect of perturbations in the infrared spectra to be studied. At the present time some research is also being conducted to study the high-resolution infrared spectra excited in emission sources.

The success of all such investigations depends upon accurate measurements of the spectral positions of infrared lines and this in turn requires adequate wavelength standards and calibration techniques suitable for the given spectrograph.

The material presented in this monograph has been chosen to help researchers meet these requirements.

Spectral charts and tables giving wavelengths and wave numbers for the emission lines of noble gases appearing in these charts are included in Chapter II. The data cover the region $1-4\,\mu$, and except for a limited number of values determined by direct interferometric measurements, the standards given are all Ritz combinations comprising calculated transitions based on energy levels derived from interferometric measurements. Many Ritz standards in addition to those given here are possible; however, this monograph includes only those that are intense enough to be observed with grating spectrographs which employ the sources and infrared detectors currently available.

The notations used for atomic spectral transitions reported in this book follow the convention presently accepted for molecular transitions according to which the upper state is written first. This does not conform with most published descriptions of atomic spectra.

Absorption standards are more widely used in the infrared because study of absorption spectra of molecules is one of the main undertakings of most infrared laboratories. Diatomic and sometimes linear polyatomic molecules possess simple structures and, therefore, wavelengths of characteristic features are suitable for use as standards. Chapter III presents data for the best available absorption standards in the near infrared. Tracings of the spectra are shown and vacuum wave numbers of spectral lines are given. In most cases, the spectra were recorded under high resolution by employing vacuum spectrographs. In a few instances in which the bands cover extended wave number intervals the spectral

v

lines observed are sketched with a uniform wave number scale. While making these sketches the resolution of the particular spectrograph employed has been preserved.

In a few of the charts presented in Chapter III, there appear absorption features other than those recommended as standards. In view of a short air path in the optics of the spectrograph and also due to impurities in the gas samples used, some of these unidentified lines result from atmospheric absorptions. It is also possible that "hot" bands of the particular molecule may explain these lines. In this monograph, no attempt was made to identify these extra lines.

As may be noted from Fig. 1 of Chapter I, in the region between 1 and $30\,\mu$, the number of available absorption standards is not adequate when gratings are used in the first order for recording spectra. However, when observations are made by using coarse gratings in higher spectral orders, the standards available do allow measurements to be made in any desired spectral region. Modern gratings ruled with interferometric controls are of such excellent quality that for spectrographs equipped with these gratings it seems possible to use the techniques of overlapping orders for determining spectral positions. Needless to say, it would be most advantageous to examine first the validity of this method by repeating measurements on known lines. In Chapter V the calibration techniques currently used in infrared laboratories are discussed.

It is hoped that this monograph will be of use to investigators who find it necessary to measure accurately spectral positions in the infrared. Although high resolving power and precision have been emphasized in this undertaking, it is believed that laboratories employing medium resolution spectrographs may find it useful to use the techniques discussed in Chapter V, and thus eliminate the necessity of basing measurements on standards which are sensitive to the spectral resolution actually used.

Infrared spectra may be divided into two kinds: the near infrared spectra which extend from 1 to $30\,\mu$, and the far infrared spectra which include wavelengths between 30 and $1000\,\mu$. Chapter IV summarizes the status of standards in the far infrared region of the spectrum. Much work remains to be done in order to establish additional standards in this region. In the spectral interval 600-50 cm^{-1} (16-200 μ), charts of the observed pure rotational lines of water vapor have been reproduced for enabling a quick location of this region.

A brief account of the data furnished in the various appendixes is given below:

Appendix I discusses the correction factor for the index of refraction of air in the case of instruments which do not operate in vacuum.

Appendix II gives a summary of the molecular constants for the bands recommended as absorption standards. In addition to providing values of these constants for use by molecular spectroscopists interested in the structural problems, the precision with which molecular constants can be determined from infrared spectroscopy at the present time is pointed out.

Appendix III is essentially a supplement to Chapter V. It contains data useful for coarse echelle-type gratings that are currently available. It is believed that the first-order wave numbers of molecular absorption standards included in this appendix will provide a useful starting point. Since the accuracy of the absorption standards has been estimated to be at least within ± 0.002 cm^{-1}, the first-order wave numbers are given to 5-decimal places to avoid rounding-off errors.

In an undertaking of this nature, the authors appearing in the various references cited have obviously contributed most significantly.

At various times during the proceedings of the Triple Commission for Spectroscopy, Dr. G. Herzberg offered many valuable comments; the authors are also grateful to him for reading the manuscript and providing valuable suggestions. The association of Mr. Edward Paul, Jr., with C. J. Humphreys, of Professor T. A. Wiggins with D. H. Rank, and that of Professor Harald H. Nielsen with K. Narahari Rao led to many fruitful results over the past several years.

Finally, the authors are grateful to Professor T. Harvey Edwards of Michigan State University and to Professor T. K. McCubbin of Pennsylvania State University for their criticisms of the manuscript.

K. Narahari Rao
Curtis J. Humphreys
D. H. Rank

CONTENTS

CHAPTER I

INTRODUCTORY REMARKS

This monograph contains a compilation of wavelength standards which are regarded as being especially suitable for use with high-resolution infrared spectrographs. Since there is wide interest in studying both emission and absorption spectra in the infrared, emission as well as absorption standards are included in this compilation. All the observational data reproduced in this book in the form of charts with identified features have been obtained by employing spectrographs equipped with plane gratings. Several of the absorption spectra in the near infrared have been recorded using spectral slit widths of about 0.05 cm^{-1} or less.

As will be apparent from an examination of the available material, some of the standards are known to a high degree of precision, but there are regions where no suitable standards are available for observations with gratings used in the first order.

In the far infrared, where there is a scarcity of standards, charts of spectra of the pure rotational lines of the water vapor molecule are furnished along with the currently available wave numbers for these rotational lines. Although these data aid primarily in the identification of the appropriate spectral regions, it is possible that, within limits, some of the single pure rotational lines of the water vapor molecule may be useful for the calibration of far infrared spectrographs.

An explanation of the charts and tables appearing in Chapters II-IV is given below. The monograph concludes with a brief discussion of some of the techniques currently employed in the determination of the positions of spectral lines recorded with high-resolution infrared spectrographs.

A. ATOMIC EMISSION STANDARDS PRESENTED IN CHAPTER II

Atomic line emission standards of argon, neon, krypton, and xenon are given in Chapter II. The wavelength region covered extends from 1 to 4 μ. The data for these standards are arranged in such a way that the table on each page gives the air wavelengths (expressed in angstroms) and vacuum wave numbers (expressed in cm^{-1}) corresponding to the spectral lines reproduced in the figure appearing on the opposite page. The indicated spectral positions (λ_{air} and ν_{vac}) also include a specification of the grating order in which the lines actually occur.

It should be mentioned that the atomic line emission standards have been considered at various times during the deliberations of Commission 14 of the International Astronomical Union which has been the recognized authority for the adoption of such standards. The

1

choice of the material presented in Chapter II of this monograph follows closely the recommendations of the Commission.

Atomic Emission Standards of Argon (pp. 12-49):

Two hundred and fourteen atomic line emission standards of argon are available for use in the region of 0.96-4.09 μ. In the region of 0.96-3.3 μ, the spectra have been recorded with a 15,000 lines per inch plane grating, whereas, for the region of 3.3-4.1 μ, a 7500 lines per inch plane grating was used. Both these gratings were ruled at the Johns Hopkins University. In the entire spectral region of 0.96-3.3 μ, a cooled PbS cell was employed to detect the infrared radiation.

All reproduced chart recordings have been obtained by rotating the grating to scan the spectra. The physical widths of the entrance and exit slits of the spectrograph were maintained at one value over wide angular stretches. The grating equation expressed in the form

$$\nu = nK \csc \theta \tag{I-1}$$

can be used as a starting point to determine the spectral slit widths corresponding to the physical slit widths. In this equation, n is the spectral order, K is the instrument constant (expressed in cm^{-1}) and θ is the angle between the central image and the spectral line located at ν cm^{-1}. From this equation, one can obtain the expression

$$d\nu = \frac{\nu(\cot \theta) S}{2f} \tag{I-2}$$

relating the physical slit width S to the spectral slit width $d\nu$ for a spectrograph of focal length f.

A 102 cm focal length Littrow-type spectrograph was employed for obtaining the information pertaining to the infrared spectrum of argon. Just beyond the visible red, the spectral slits were calculated to be 2 cm^{-1}, whereas at 3.3 μ they were about 0.05 cm^{-1}. Between 3.3-4.1 μ, where the 7500 lines per inch grating was used, the calculated value of the spectral slit width came close to 0.2 cm^{-1}.

The source[1] in which the argon lines were excited was a sealed section of vycor or quartz tube of about 12 cm in length and 6 mm in diameter, filled with argon. It was excited by a 2450 Mc "microtherm" oscillator[2].

The spectral positions (ν_{vac}) of argon quoted in the tables are values calculated by employing the Ritz combination principle. The level values introduced into these calculations are derived from extremely precise interferometric measurements, substantiated by independent, concordant observations. In view of the precision with which some of the low

---

[1] C. J. Humphreys and E. Paul, Jr., *Appl. Opt.* **2**, 691(1963).
[2] The particular one used was procured from the Raytheon Co., Power Equipment Dept., Richards Avenue, South Norwalk, Connecticut.

energy levels of argon are known[3,4], use of the Ritz combination principle to determine the transitions between these levels seems to be a satisfactory procedure to adopt. In fact, a few of the infrared lines of argon have been measured by three different investigators employing entirely independent techniques, one of them being a direct intercomparison with the primary standard of length, the Kr[86] line with λ_{vac} = 6057.80210$_5$ Å. Table I shows a comparison of the results obtained by the three investigators (columns 1-3) along with the values derived from the Ritz combination principle (column 4). The agreement between the data appearing in columns 1-4 is excellent. The range of differences between the wave number values obtained by different authors for the same line corresponds to about ±0.002 Å or better.

TABLE I

Vacuum Wave Numbers (cm⁻¹) of Ar I Lines in the Infrared

Humphreys and Paul[a]	Littlefield and Rowley[a]	Peck[a]	Ritz combination principle	
			ν_{vac} (cm⁻¹)	λ_{air} (Å)[b]
7808.693	7808.689	7808.693	7808.694	12802.737
7715.929	7715.929	7715.930	7715.929	12956.658
7403.084	7403.085	7403.084	7403.085	13504.190
7338.705	7338.700	7338.702	7338.704	13622.659
7287.394	7287.393	7287.394	7287.393	13718.577
5901.373	5901.372	5901.373	5901.372	16940.584

[a] G. Herzberg, *Trans. Intern. Astron. Union*, **11A**, 97-117 (1962).

[b] This column of λ_{air} (Å) is given for the sake of illustrative examples worked out in Appendix I.

Reduction of spectral positions to their vacuum values was made by employing Edlén's formula[5] for the dispersion of air. Based on this work of Edlén, publications appeared[6,7] which would allow one to convert wavelengths in air to wave numbers in vacuum. However, Dr. Edlén has recently prepared tables suitable for use with desk calculators. Conversion of wavelengths in air to wave numbers in vacuum and vice versa can be made easily with the aid of these tables. Through the courtesy of Dr. Edlén, these tables are reproduced in Appendix I, and a few examples have been worked to illustrate how they should be used. Infrared spectrographs operated in air are sometimes convenient to use and, in fact, are

[3] B. Edlén *Trans. Intern. Astron. Union* **9**, 201-227 (1957).
[4] C. M. Sitterly, *Trans. Intern. Astron. Union* **12** (to be published).
[5] B. Edlén, *J. Opt. Soc. Am.* **43**, 339 (1953).
[6] C. D. Coleman, W. R. Bozman, and W. F. Meggers, *Natl. Bur. Std., (U.S.) Monograph* **3** (1960).
[7] R. Penndorf, *J. Opt. Soc. Am.* **47**, 176 (1957).

being used in some laboratories. Therefore, it is important to have access to one or the other of these sources to permit the determination of vacuum values for spectral positions.

The relative intensities indicated in the argon spectral charts are on an approximate logarithmic scale. It is hoped that they will aid in determining the amplifier and other settings required for the observation of the infrared argon lines. Much work remains to be done in developing suitable sources for exciting the infrared spectrum of argon. Many other argon transitions are located in the infrared region. A list of argon lines calculated as Ritzian combinations has been published by Hymphreys[8], including a notation indicating which lines were actually observed with the source excited by the 2450 Mc "microtherm" oscillator.

Atomic Emission Standards of Neon (pp. 50-51):

By employing a discharge tube similar to that used for argon, it is possible to observe some neon lines in the infrared. The infrared spectrum of neon is somewhat weaker than that of argon and only about 30-40 lines have been observed using high-resolution infrared spectrographs. The chart on page 50 shows a tracing[9] of the neon lines observed in the region of 2.0-$2.6\,\mu$. They were recorded with a 1 m focal length Pfund-type spectrograph equipped with a 15,000 lines per inch plane grating ruled by Bausch and Lomb. A liquid nitrogen-cooled PbS detector was employed. The spectra reproduced were recorded by using spectral slit widths [calculated from Eq. (I-2)] of nearly 1 cm^{-1}. Observations of all the neon lines were made by keeping the amplifier at one gain setting; and therefore, it is believed that the heights of the lines indicated in the chart provide estimates of the relative intensities of the observed neon lines.

Again, spectral positions listed have been derived by employing the Ritz combination principle. With one exception (λ_{air} 24452.415 Å, $3s_3$-$3p_{10}$) the three-place entries are $4d$-$3p$ transitions. Transitions in the category $3p$-$2s$ are given to four decimal places in both wavelengths and wave numbers. Four-place $3p$ energy levels were adopted[10] in the year 1957, and four-place $2s$ energy levels will appear in a forthcoming issue[11] of the *Transactions of the International Astronomical Union*.

Atomic Emission Standards of Krypton (pp. 52-67):

The first spectrum of krypton, Kr I, has been utilized less frequently as a source of standard wavelengths than the spectrum of Ar I. The principal reason for this preference is that argon, which in its natural composition contains 99.6% of the isotope of mass number 40, is expected to provide symmetrical line shapes. Another reason is that a sufficient

---

[8] C. J. Humphreys, *Appl. Opt.* **2**, 1155 (1963).
[9] M. E. Mickelson, The Ohio State University, private communication (1965).
[10] See Edlén[3].
[11] See Sitterly[4].

number of highly precise wavelength determinations have been made for argon, which are
both independent and concordant, and permit the adoption of such wavelengths as standards.

For krypton, the requirement of symmetrical line shapes has been met recently by
the availability of the isotope of even-numbered mass Kr^{86} in adequate quantities for filling
sources. In fact, the line of Kr^{86} having a wavelength in standard air (dry air containing
0.03% CO_2 by volume at normal pressure of 760 mm Hg and a temperature of 15°C) of
6056.12525 Å has been adopted by the Eleventh General Conference of Weights and Measures[12] as the primary standard of length, superseding the material standard in the form of
a platinum iridium bar that has long been maintained at the International Bureau of Weights
and Measures. But work on the determinations of wavelengths of the highest accuracy of
other Kr lines was started too recently to lead to the accumulation of a sufficient number of
independent, concordant measurements. Therefore, level values have not yet been adopted
from which standards may be calculated as Ritzian combinations.

The wavelengths of the very intense $2s$-$2p$ and $3d$-$2p$ transitions in Kr^{86} I which occur
mostly in the interval between 1.1 and 2.2 μ, have been measured interferometrically by
Humphreys and Paul[13] and by Littlefield and Sharp[14]. The results of the first named set of
determinations are included in Chapter II. The principal deficiency in the interferometric
observation of the Kr spectrum is the almost complete lack of measurements of the patterns
originating in the transitions $2p$-$1s$. These occur for the most part in a photographically
accessible region between 0.7 and 0.9 μ. When such measurements have been made and
the $2p$ levels determined, the positions of all the intense infrared lines in Kr^{86} may be computed.

In spite of the observable isotope structure accompanying some of the very intense
lines, natural krypton does provide a very useful set of standards. The naturally occurring
element consists of a mixture of six isotopes, five of which are moderately abundant. Only
one of these, Kr^{83}, occurring in the proportion of 11.55% is of odd-numbered atomic mass.
The relatively faint hyperfine structure originates in Kr^{83}. The predominant line intensity
is accounted for by the unresolved complex of components originating in the even-numbered
isotopes, and the measured position of a given line is almost the same as that of the corresponding line of Kr^{84}, the most abundant isotope.

Although the wavelengths of several groups of lines in this spectrum have been measured several times over a period of many years, there is an insufficient number of independent determinations of wavelengths of infrared lines, either in the photographically accessible region or in the region of radiometric exploration, to permit adoption as standards.
The most recent interferometric infrared measurements in the photographic infrared region

[12] See, for instance, G. Herzberg, *Trans. Intern. Astron. Union* **11B**, 208-210 (1962).
[13] See Sitterly[4].
[14] See Sitterly[4].

are now more than 30 years old, having been made by Meggers and Humphreys[15] in the year 1934. The intense $3d$-$2p$ and $2s$-$2p$ combinations were observed interferometrically by Humphreys and Paul[16] in 1961. It appears improbable that interferometric measurements on the naturally occurring element will be continued. Future determinations of wavelengths in krypton are expected to make use of the available isotopes, Kr[84] or Kr[86].

A description of the arrangement of the tables of Kr standards is given below:

Wavelengths and wave numbers of selected lines in the spectra of natural krypton and of Kr[86] are given under appropriate headings. In the case of natural krypton and of Kr[86], the four-decimal place entries designate the direct interferometric determinations made by Humphreys and Paul[17]. The three-decimal place entries are positions of calculated Ritzian combinations, making use of the $2s$ and $3d$ level values from the work of Humphreys and Paul just mentioned, the $1s$ and $2p$ level values from the 1934 work of Meggers and Humphreys[18]. Two-place entries are calculated from levels as given in "Atomic Energy Levels"[19].

The column designated "Grating order" shows the orders as they appear in the section of record chart selected for observation. A plane grating with 7500 lines per inch on its surface (ruled at the Johns Hopkins University), was employed as the dispersing element, and a liquid nitrogen-cooled PbS detector was used for detecting the infrared signal. The intensities are relative on an arbitrary scale as they appear on the selected record. No account is taken of the blaze characteristics of the grating or of the spectral response of the detector. The spectral slit widths employed were calculated to be about 0.1 cm^{-1}. This estimate of the spectral slit width corresponds to the wave number in the first order of the grating. Since the same physical slit width was used throughout the spectral region scanned, the effective spectral slit width does not have the same value throughout.

Atomic Emission Standards of Xenon (pp. 68-83):

Xenon in its naturally occurring form is composed of nine isotopes of which seven are present in significant amounts. Two of these, of odd mass numbers 129 and 131, are in concentrations of 26.44 and 21.18%, respectively, making up approximately half of the atomic mass. On this account, the hyperfine structures are strongly developed, causing the lines to appear broadened under ordinary conditions of observation. Natural xenon is therefore generally considered to be the least useful of the noble gases as a source of standard wavelengths. Many of the lines have been observed interferometrically and show satisfactorily reproducible wavelengths. The possibility of making use of these wavelengths

[15] W. F. Meggers, and C. J. Humphreys, *J. Res. Natl. Bur. Std.* **13**, 293 (1934); RP 710.
[16] See Sitterly[4].
[17] See Sitterly[4].
[18] See Meggers and Humphreys[15].
[19] C. E. Moore, *Natl. Bur. Std. (U. S.), Circ.* **467**, Vol. II (1952).

for comparison purposes in regions deficient in other standards should not be dismissed.

The large atomic mass of the element, which has the effect of minimizing Doppler broadening, makes the prospect of utilizing an isotope of xenon of even mass number extremely attractive. In fact, from the standpoint of sharpness, the lines of such an isotope should be superior to those of any other noble gas. The isotope of mass number 136 has been enriched to the point of very high purity at the Physikalisch-Technische Bundesanstalt (P. T. B.) of the West German Republic in a program directed by Dr. E. Englehard. Sources or source tube fillings have been made available by that institution during recent years.

Making use of Xe^{136} sources designed for microwave excitation and filled at the P. T. B., Humphreys and Paul[20] made interferometric measurements of ten lines of this isotope. These lines, located in the interval between 2.0 and 3.5 μ, consist mostly of the very intense $3d$-$2p$ combinations. They are expected to provide the most useful atomic line emission standards for this region.

The wave numbers of the lines of natural xenon are computed from the levels listed in "Atomic Energy Levels"[21]. Three-decimal place values are retained in instances where both levels involved in a transition have been given to three places. Such levels have been derived from interferometric measurements. The wavelengths of the lines of Xe^{136} carried to four decimal places are results of direct interferometric measurements. The relative intensities are on an arbitrary scale appropriate to the chart section selected for illustration; the same considerations apply as in the instance of the corresponding tables for Kr I. The grating used for recording the Xe spectra is the same as that used for the Kr lines. The spectral slits employed are calculated to be about 0.3 cm^{-1}. Again, this value is variable in the region studied for the Xe spectra because the same physical slit width was maintained throughout the scans. It should be noted that the calculation of the spectral slit width corresponds to the wave number in the first order of the grating.

B.   MOLECULAR ABSORPTION STANDARDS BETWEEN 1.5 and 16 μ PRESENTED IN CHAPTER III

Since much of the work done in infrared spectroscopy pertains to study of the absorption spectra of polyatomic molecules, it is entirely logical to think in terms of absorption standards. In 1950, use of the rotational lines of the vibration rotation bands of the carbon monoxide molecule as secondary standards in the infrared was proposed by Rao[22] as a result of studies on some of the electronic bands of CO occurring in the ultraviolet and the 4-0 vibration rotation band of CO occurring in the photographic infrared. The absorption

[20] See Sitterly [4].
[21] See Moore [19].
[22] K. Narahari Rao, *J. Chem. Phys.* **18**, 213 (1950).

spectrum of CO in the infrared is simple and the spacings between the rotational structure (3-4 cm^{-1}) are convenient for use with high-resolution infrared spectrographs. Also, the 1-0 and 2-0 vibration rotation bands of CO can be observed by employing relatively small absorption paths. In the year 1962, it was shown by Rao and his co-workers[23] that the infrared CO bands are even more useful as standards when one makes use of coarse gratings for studying high-resolution infrared spectra. In particular, the work of Rao *et al.* demonstrated that a coarse grating, for instance one with 40 grooves/mm (1000 lines per inch) used echelle fashion allows us to refer the wavelength measurements in the 1-40 μ region to the positions of the rotational lines of the 1-0 and 2-0 bands of the CO molecule occurring, at 4.66 and 2.33 μ respectively. The experience of McCubbin *et al.*[24] is similar in this respect. They found that use of a coarse echelle with 30 grooves/mm for high resolution studies in the infrared is even more satisfactory for employing the rotational lines of the vibration rotation bands of CO as absorption standards.

Recently, some of the rotational lines of the vibration rotation bands of the CO, HCN, and N_2O molecules have been measured with a high precision by Rank and his co-workers. Two rotational lines in the bands of CO and HCN have been measured interferometrically[25] relative to the green line of Hg^{198}. They are the R(18) line of the 2-0 band of CO, and the P(1) line of the $2\nu_3$ band of HCN. All the other lines in the 2-0 band of CO and the $2\nu_3$ band of HCN have been measured interferometrically relative to these two lines. Other absorption standards quoted in Chapter III have been measured[26, 27] by using echelle gratings and overlapping orders of the above CO and HCN bands.

In all these molecular standards, in a particular band, there is excellent accuracy for the spectral positions of the rotational lines relative to one another. In the case of the absorption standards, the tables list not only the "observed" values but also those labeled "calculated" values. These "calculated" values were obtained in the following way. First, the observed vibration rotation bands were analyzed and the transitions involved were identified. The measured spectral positions were then used to evaluate the molecular constants (the molecular constants involved are summarized in Appendix II) for both the upper and lower states of a transition by employing the techniques of combination relations[28]. The theoretical formulas which express the wave numbers of the rotational lines of a vibration rotation band as a function of the molecular constants are also well known[29]. It is a straight-

[23] K. Narahari Rao, W. W. Brim, V. L. Sinnett, and R. H. Wilson, *J. Opt. Soc. Am.* **52**, 862 (1962).

[24] T. K. McCubbin, Jr., J. A. Lowenthal, and H. R. Gordon, *Appl. Opt.* **6**, 711 (1965).

[25] D. H. Rank, D. P. Eastman, B. S. Rao, and T. A. Wiggins, *J. Opt. Soc. Am.* **51**, 929 (1961).

[26] See Rank *et al.*[25].

[27] W. W. Brim, J. M. Hoffman, H. H. Nielsen, and K. Narahari Rao, *J. Opt. Soc. Am.* **50**, 1208 (1960).

[28] See, for example, G. Herzberg, "Spectra of Diatomic Molecules." Van Nostrand, Princeton, New Jersey, 1950.

[29] See, for example, Herzberg[28].

forward procedure to arrive at "calculated" values for the wave numbers of the rotational lines of the vibration rotation bands by inserting the molecular constants determined above in the appropriate theoretical formulas. Therefore, these "calculated" values are, in a sense, "smoothed out" values. It would obviously be more advantageous to use the "calculated" values for calibration purposes.

The accuracy of the absorption standards is estimated to be ± 0.002 cm^{-1}. Only vacuum wave numbers (expressed in cm^{-1}) of the absorption standards are given in this monograph. In the spectrographs employed for the measurement of the infrared absorption spectra of CO, HCN, and N_2O, the grating was always kept in vacuum.

It may be recalled that the figures shown in Chapter III were obtained by making use of high-resolution infrared spectrographs. These instruments are equipped with modern ruled echelles having a ruled area of 10×5 in. The "masters" for these echelles were ruled by Professor George Harrison at the Massachetts Institute of Technology, and the replicas were prepared by Bausch and Lomb. It is imperative that a comparison be made with these spectral charts before making use of the standards, because in the case of molecular bands there can always be other overlying bands, and one should be careful to avoid using blended lines.

C.   MOLECULAR ABSORPTION STANDARDS IN THE FAR INFRARED AT WAVE-
     LENGTHS LONGER THAN 16 μ PRESENTED IN CHAPTER IV

In the far infrared region, the standards available are not entirely satisfactory. The 116 pure rotational lines of H_2O between 550 cm^{-1} and 271 cm^{-1} are believed to be known[30] to an accuracy of ± 0.03 cm^{-1} for unblended lines. The spectra sketched for these lines of the H_2O molecule (p.144) were based upon the observational data obtained by employing a 1000 lines per inch Bausch and Lomb plane grating in a 1 m focal length Pfund-type vacuum infrared spectrograph.

The H_2O lines between 271 cm^{-1} and 120 cm^{-1} are determined purely from energy levels[31], and therefore, until verified by direct measurement, they should be used only for approximate calibration work requiring an accuracy of ± 0.05 to ± 0.1 cm^{-1}. Maps are provided (pp. 146-147) for the pure rotational lines of H_2O in the region 271-120 cm^{-1}. Wave numbers of the lines are included in these maps. The observational data were obtained with a Perkin-Elmer Model 301 spectrophotometer.

At wavelengths longer than 83.3 μ (120 cm^{-1}) the pure rotational lines of the CO, HCN, and N_2O molecules can be considered to be known to a very high degree of precision, because the ground state rotational constants for these molecules have been very well established and the spectral positions quoted in the tables are values "calculated" by using the standard theory.

[30] See Rao *et al*. [23].
[31] K. Narahari Rao, R. V. deVore, and E. K. Plyler, *J. Res. Natl. Bur. Std.* **67A**, 351 (1963).

D. SUMMARY OF AVAILABLE ABSORPTION STANDARDS IN THE INFRARED

The available absorption standards in the region 1-1000 μ are summarized in Fig. 1.

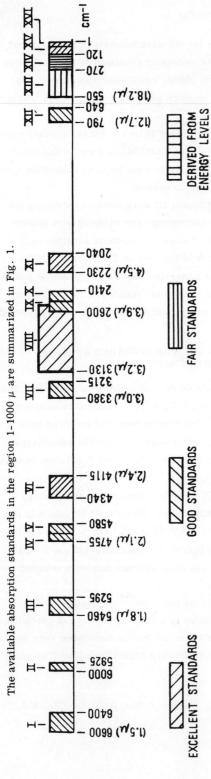

Fig. 1. Summary of provisional absorption standards in the infrared.

Serial number as in Fig. 1	Number of absorption standards and molecule responsible	Average separation between consecutive absorption lines (cm^{-1})	Serial number as in Fig. 1	Number of absorption standards and molecule responsible	Average separation between consecutive absorption lines (cm^{-1})
I	66($HC^{12}N^{14}$)	2.9	IX	114($N_2^{14}O^{16}$)	0.8
II	81($N_2^{14}O^{16}$)	0.7	X	113($N_2^{14}O^{16}$)	0.8
III	57($HC^{12}N^{14}$)	2.9	XI	51($C^{12}O^{16}$)	3.8
IV	110($N_2^{14}O^{16}$)	0.7	XII	50($HC^{12}N^{14}$)	3.0
V	96($N_2^{14}O^{16}$)	0.8	XIII	116(H_2O^{16})	Irregular
VI	61($C^{12}O^{16}$)	3.8	XIV	74(H_2O^{16})	Irregular
VII	55($HC^{12}N^{14}$)	3.0	XV	30($C^{12}O^{16}$) and 40($HC^{12}N^{14}$)	3.8 and 2.9
VIII	31(HCl^{35})	9.8 at R(15) and 28.0 at P(15)	XVI	50($N_2^{14}O^{16}$)	0.8

EXCELLENT STANDARDS

GOOD STANDARDS

FAIR STANDARDS

DERIVED FROM ENERGY LEVELS

CHAPTER II
Infrared Emission Standards

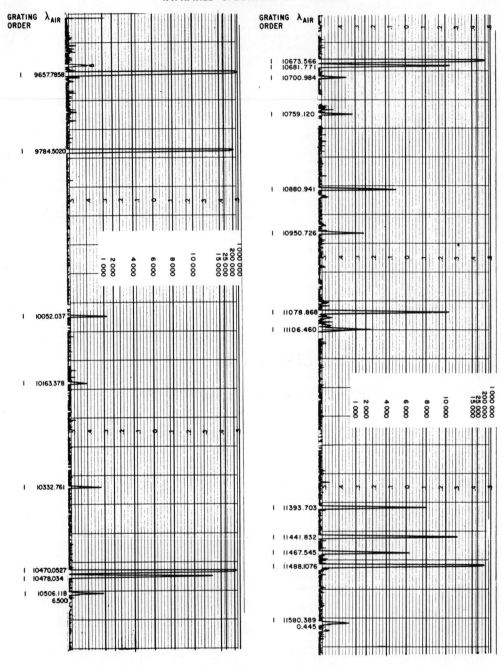

Wavelengths (Air) and Vacuum Wave Numbers (cm⁻¹) of the Near
Infrared Emission Lines of Argon[a]

Grating order	λ_{air} (Å)	ν_{vac} (cm⁻¹)
1	9657.7858	10351.5012
1	9784.5020	10217.4424
1	10052.037	9945.506
1	10163.378	9836.552
1	10332.761	9675.304
1	10470.0527	9548.4338
1	10478.034	9541.161
1	10506.118	9515.656
1	10506.500	9515.310
1	10673.566	9366.374
1	10681.771	9359.179
1	10700.984	9342.375
1	10759.120	9291.895
1	10880.941	9187.865
1	10950.726	9129.314
1	11078.868	9023.722
1	11106.460	9001.304
1	11393.703	8774.375
1	11441.832	8737.467
1	11467.545	8717.875
1	11488.1076	8702.2714
1	11580.389	8632.925
1	11580.445	8632.883

[a] The four-decimal place accuracy quoted for the wave numbers
of the two spectral lines with λ9657.7858 and λ9784.5020 is based on
the precision available for the energy levels involved in their transitions [B. Edlén, *Trans. Intern. Astron. Union* **9**, pp. 201-227 (1957)].
    Recent additional measurements justify quoting the wave num-
bers of the lines λ10470.0527 and λ11488.1076 to the fourth decimal
place.

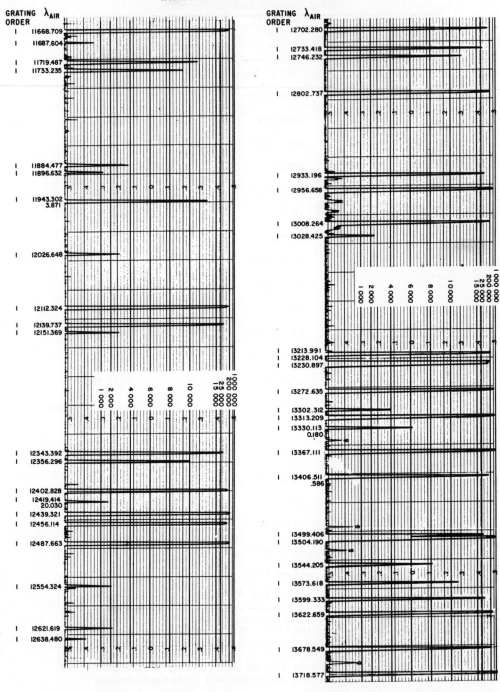

14

Wavelengths (Air) and Vacuum Wave Numbers (cm⁻¹) of the
Near Infrared Emission Lines of Argon (Continued)

Grating order	λ_{air} (Å)	ν_{vac} (cm⁻¹)	Grating order	λ_{air} (Å)	ν_{vac} (cm⁻¹)
1	11668.709	8567.583	1	12746.232	7843.310
1	11687.604	8553.732	1	12802.737	7808.694
1	11719.487	8530.462	1	12933.196	7729.926
1	11733.235	8520.466	1	12956.658	7715.929
1	11884.477	8412.035	1	13008.264	7685.319
1	11896.632	8403.440	1	13028.425	7673.426
1	11943.302	8370.603	1	13213.991	7565.667
1	11943.871	8370.204	1	13228.104	7557.595
1	12026.648	8312.594	1	13230.897	7556.000
1	12112.324	8253.795	1	13272.635	7532.239
1	12139.737	8235.157	1	13302.312	7515.435
1	12151.369	8227.274	1	13313.209	7509.283
1	12343.392	8099.285	1	13330.113	7499.761
1	12356.296	8090.826	1	13330.180	7499.723
1	12402.828	8060.472	1	13367.111	7479.003
1	12419.414	8049.707	1	13406.513	7457.022
1	12420.030	8049.308	1	13406.586	7456.981
1	12439.321	8036.825	1	13499.406	7405.708
1	12456.114	8025.990	1	13504.190	7403.085
1	12487.663	8005.713	1	13544.205	7381.213
1	12554.324	7963.204	1	13573.618	7365.218
1	12621.619	7920.747	1	13599.333	7351.292
1	12638.480	7910.180	1	13622.659	7338.704
1	12702.280	7870.449	1	13678.549	7308.718
1	12733.418	7851.203	1	13718.577	7287.393

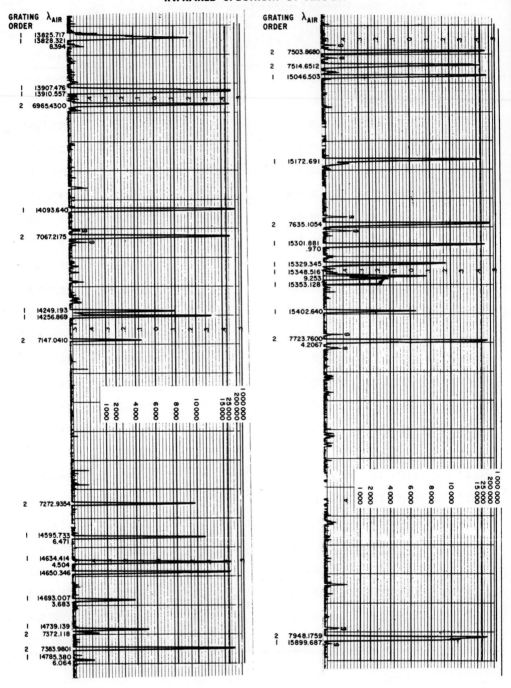

Wavelengths (Air) and Vacuum Wave Numbers (cm^{-1}) of the Near Infrared Emission Lines of Argon (continued)[a]

Grating order	λ air (Å)	ν vac (cm^{-1})	Grating order	λ air (Å)	ν vac (cm^{-1})
1	13825.717	7230.921	1	14693.683	6803.786
1	13828.321	7229.559	1	14739.139	6782.803
1	13828.394	7229.521	1	14785.380	6761.590
1	13907.476	7188.412	1	14786.064	6761.277
1	13910.557	7186.820	1	15046.503	6644.246
1	14093.640	7093.460	1	15172.691	6588.988
1	14249.193	7016.023	1	15301.881	6533.359
1	14256.869	7012.246	1	15301.970	6533.321
1	14595.733	6849.445	1	15329.345	6521.654
1	14596.471	6849.099	1	15348.516	6513.508
1	14634.414	6831.341	1	15349.253	6513.195
1	14634.504	6831.299	1	15353.128	6511.551
1	14650.346	6823.912	1	15402.640	6490.620
1	14693.007	6804.099	1	15899.687	6287.714
2	6965.4300	14352.6566	2	7503.8680	13322.7918
2	7067.2175	14145.9401	2	7514.6512	13303.6742
2	7147.0410	13987.9486	2	7635.1054	13093.7918
2	7272.9354	13745.8188	2	7723.7600	12943.4998
2	7372.118	13560.887	2	7724.2067	12942.7514
2	7383.9801	13539.1023	2	7948.1759	12578.0434

[a] The spectral lines observed in the second order of the grating are mosly transitions between the levels 2p-1s. Argon lines in this category are International Secondary Standards. The justification for the retention of four decimal places in the wave numbers is based on the adoption of the level values to four-place accuracy by the International Astronomical Union. [B. Edlén, *Trans. Intern. Astron. Union* 9, pp. 201-227 (1957)]. The line with λ7372.118 results from the transition 4d-2p.

INFRARED SPECTRUM OF ARGON

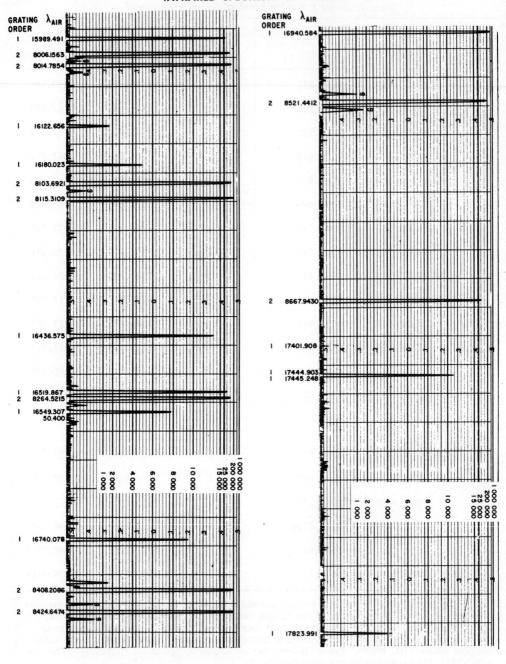

Wavelengths (Air) and Vacuum Wave Numbers (cm⁻¹) of the Near
Infrared Emission Lines of Argon (Continued)

Grating order	λ_{air} (Å)	ν_{vac} (cm⁻¹)
1	15989.491	6252.399
1	16122.656	6200.758
1	16180.023	6178.773
1	16436.575	6082.331
1	16519.867	6051.664
1	16549.307	6040.899
1	16550.400	6040.500
1	16740.078	5972.056
1	16940.584	5901.372
1	17401.908	5744.927
1	17444.903	5730.768
1	17445.248	5730.655
1	17823.991	5608.884
2	8006.1563	12486.9540
2	8014.7854	12473.5100
2	8103.6921	12336.6620
2	8115.3109	12318.9996
2	8264.5215	12096.5890
2	8408.2086	11889.8725
2	8424.6474	11866.6722
2	8521.4412	11731.8810
2	8667.9430	11533.5946

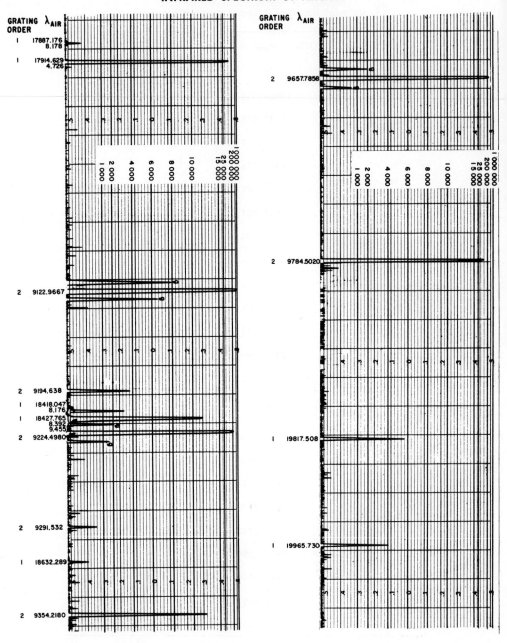

Grating order	λ_{air} (Å)	ν_{vac} (cm⁻¹)
1	17887.176	5589.071
1	17888.178	5588.758
1	17914.629	5580.506
1	17914.726	5580.476
1	18418.047	5427.975
1	18418.176	5427.937
1	18427.765	5425.112
1	18428.392	5424.928
1	18429.455	5424.615
1	18632.289	5365.562
1	19817.508	5044.666
1	19965.730	5007.215
2	9122.9667	10958.3390
2	9194.638	10872.920
2	9224.4980	10837.7242
2	9291.532	10759.536
2	9354.2180	10687.4322
2	9657.7858	10351.5012
2	9784.5020	10217.4424

INFRARED SPECTRUM OF ARGON

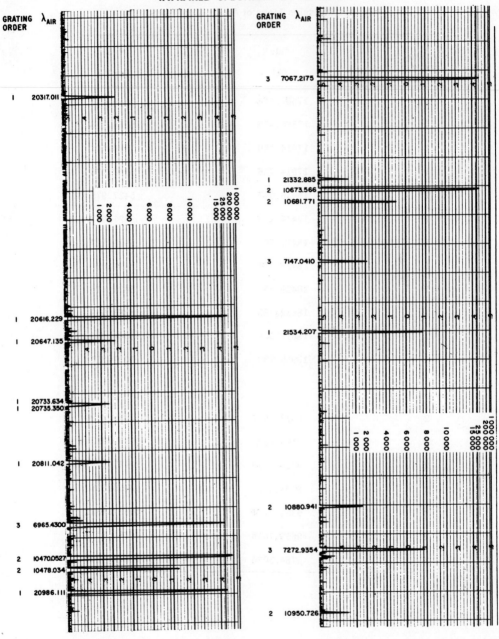

Wavelengths (Air) and Vacuum Wave Numbers (cm⁻¹) of the
Near Infrared Emission Lines of Argon (Continued)

Grating order	λ_{air} (Å)	ν_{vac} (cm⁻¹)	Grating order	λ_{air} (Å)	ν_{vac} (cm⁻¹)
1	20317.011	4920.640	2	10470.0527	9548.4338
1	20616.229	4849.224	2	10478.034	9541.161
1	20647.135	4841.966	2	10673.566	9366.374
1	20733.634	4821.765	2	10681.771	9359.179
1	20735.350	4821.366	2	10880.941	9187.865
1	20811.042	4803.830	2	10950.726	9129.314
1	20986.111	4763.756			
1	21332.885	4686.319			
1	21534.207	4642.507			
3	6965.4300	14352.6566			
3	7067.2175	14145.9401			
3	7147.0410	13987.9486			
3	7272.9354	13745.8188			

INFRARED SPECTRUM OF ARGON

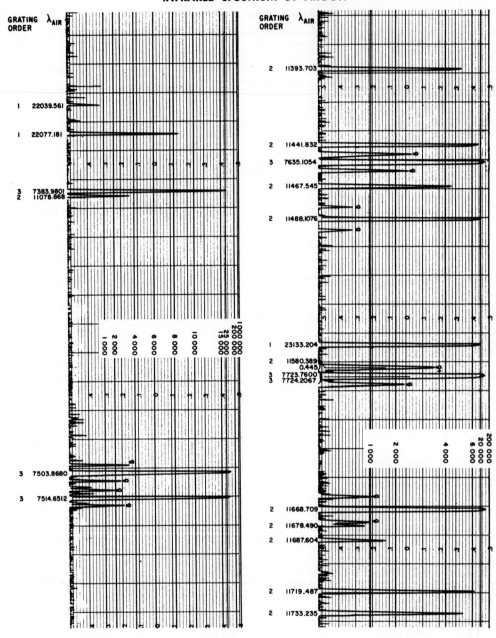

Wavelengths (Air) and Vacuum Wave Numbers (cm⁻¹) of the Near Infrared Emission Lines of Argon (Continued)

Grating order	λ_{air} (Å)	ν_{vac} (cm⁻¹)	Grating order	λ_{air} (Å)	ν_{vac} (cm⁻¹)
1	22039.561	4536.057	2	11078.868	9023.722
1	22077.181	4528.328	2	11393.703	8774.375
1	23133.204	4321.611	2	11441.832	8737.467
			2	11467.545	8717.875
			2	11488.1076	8702.271
			2	11580.389	8632.925
			2	11580.445	8632.883
			2	11668.709	8567.583
			2	11678.490	8560.407
			2	11687.604	8553.732
			2	11719.487	8530.462
			2	11733.235	8520.466
3	7383.9801	13539.1023			
3	7503.8680	13322.7918			
3	7514.6512	13303.6742			
3	7635.1054	13093.7918			
3	7723.7600	12943.4998			
3	7724.2067	12942.7514			

INFRARED SPECTRUM OF ARGON

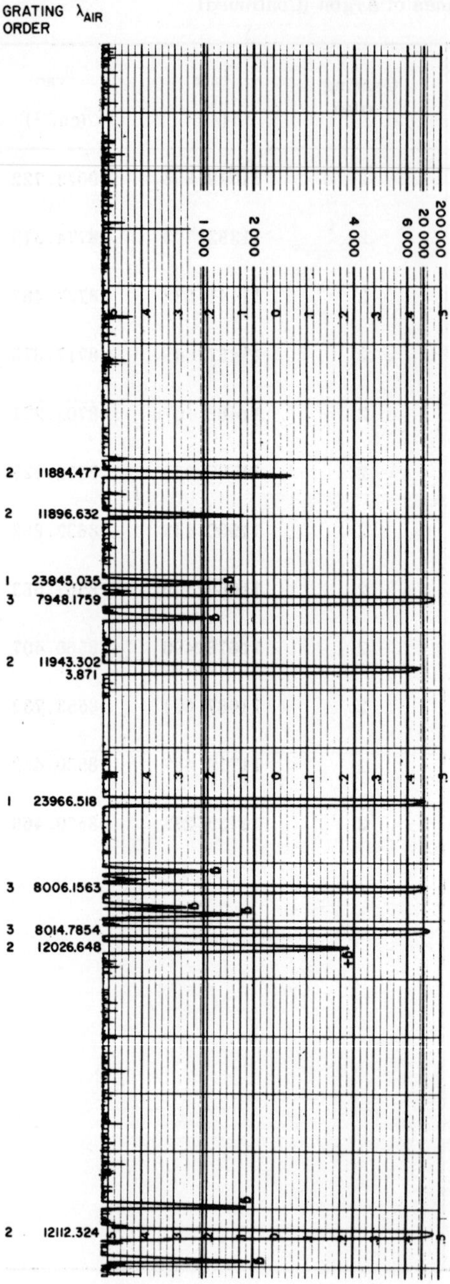

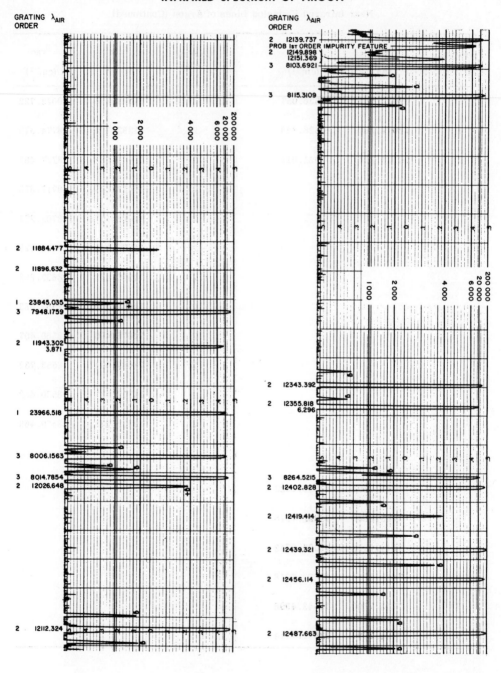

GRATING ORDER
λAIR
2 11884.477
2 11896.632
1 23845.035
3 7948.1759
2 11943.302
3.871
1 23966.518
3 8006.1563
3 8014.7854
2 12026.648
2 12112.324
1000
2000
4000
6000
20000
200000

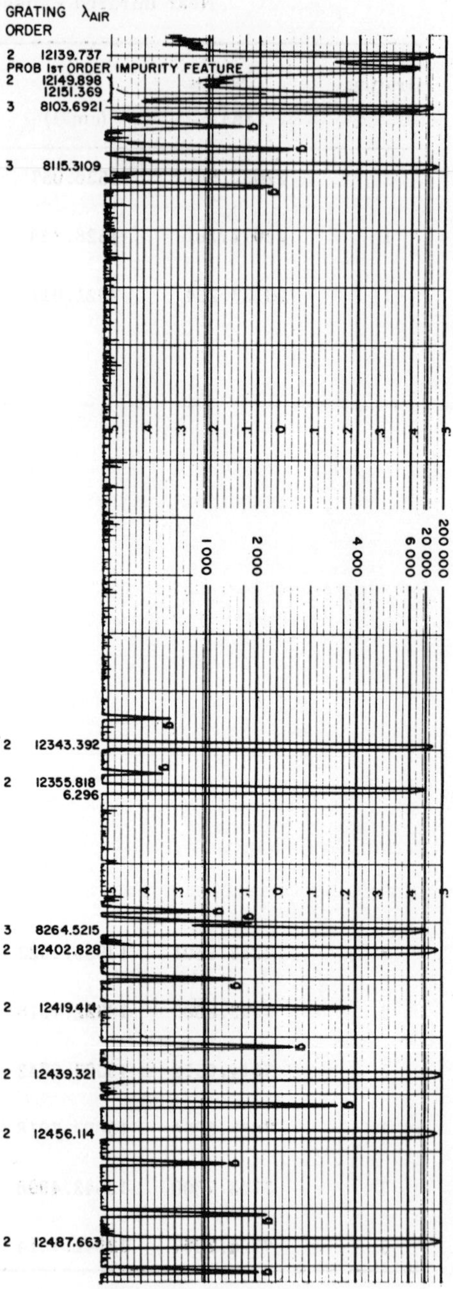

GRATING ORDER
λAIR
2 12139.737
PROB 1st ORDER IMPURITY FEATURE
2 12149.898
12151.369
3 8103.6921
3 8115.3109
2 12343.392
2 12355.818
6.296
3 8264.5215
2 12402.828
2 12419.414
2 12439.321
2 12456.114
2 12487.663
1000
2000
4000
6000
20000
200000

Wavelengths (Air) and Vacuum Wave Numbers (cm⁻¹) of the
Near Infrared Emission Lines of Argon (Continued)

Grating order	λ_{air} (Å)	ν_{vac} (cm⁻¹)	Grating order	λ_{air} (Å)	ν_{vac} (cm⁻¹)
1	23845.035	4192.601	2	11884.477	8412.035
1	23966.518	4171.349	2	11896.632	8403.440
			2	11943.302	8370.603
			2	11943.871	8370.204
			2	12026.648	8312.594
			2	12112.324	8253.795
			2	12139.737	8235.157
			2	12149.898	8228.270
			2	12151.369	8227.274
			2	12343.392	8099.285
			2	12355.818	8091.139
			2	12356.296	8090.826
			2	12402.828	8060.472
			2	12419.414	8049.707
			2	12439.321	8036.825
			2	12456.114	8025.990
			2	12487.663	8005.713
3	7948.1759	12578.0434			
3	8006.1563	12486.9540			
3	8014.7854	12473.5100			
3	8103.6921	12336.6620			
3	8115.3109	12318.9996			
3	8264.5215	12096.5890			

INFRARED SPECTRUM OF ARGON

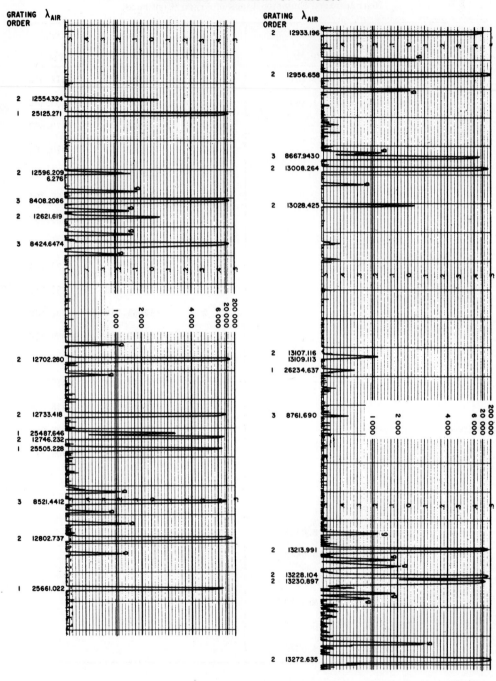

Wavelengths (Air) and Vacuum Wave Numbers (cm⁻¹) of the
Near Infrared Emission Lines of Argon (Continued)

Grating order	λ_{air} (Å)	ν_{vac} (cm⁻¹)	Grating order	λ_{air} (Å)	ν_{vac} (cm⁻¹)
1	25125.271	3978.971	2	12554.324	7963.204
1	25487.646	3922.399	2	12596.209	7936.725
1	25505.228	3919.695	2	12596.276	7936.683
1	25661.022	3895.898	2	12621.619	7920.747
1	26234.637	3810.715	2	12702.280	7870.449
1	26543.041	3766.439	2	12733.418	7851.203
			2	12746.232	7843.310
			2	12802.737	7808.694
			2	12933.196	7729.926
			2	12956.658	7715.929
			2	13008.264	7685.319
			2	13028.425	7673.426
			2	13107.116	7627.357
			2	13109.113	7626.195
			2	13213.991	7565.667
			2	13228.104	7557.595
			2	13230.897	7556.000
			2	13272.635	7532.239
3	8408.2086	11889.8725			
3	8424.6474	11866.6722			
3	8521.4412	11731.8810			
3	8667.9430	11533.5946			
3	8761.690	11410.190			

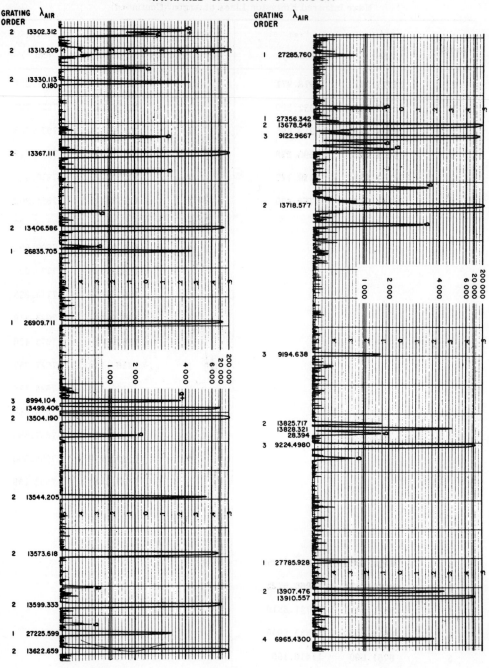

Wavelengths (Air) and Vacuum Wave Numbers (cm⁻¹) of the
Near Infrared Emission Lines of Argon (Continued)

Grating order	λ_{air} (Å)	ν_{vac} (cm⁻¹)	Grating order	λ_{air} (Å)	ν_{vac} (cm⁻¹)
1	26835.705	3725.363	2	13302.312	7515.435
1	26909.711	3715.117	2	13313.209	7509.283
1	27225.599	3672.012	2	13330.113	7499.761
1	27285.760	3663.916	2	13330.180	7499.723
1	27356.342	3654.463	2	13367.111	7479.003
1	27785.928	3597.963	2	13406.586	7456.981
			2	13499.406	7405.708
			2	13504.190	7403.085
			2	13544.205	7381.213
			2	13573.618	7365.218
			2	13599.333	7351.292
			2	13622.659	7338.704
			2	13678.549	7308.718
			2	13718.577	7287.393
			2	13825.717	7230.921
			2	13828.321	7229.559
			2	13828.394	7229.521
			2	13907.476	7188.412
			2	13910.557	7186.820
3	8994.104	11115.344	4	6965.4300	14352.6566
3	9122.9667	10958.3390			
3	9194.638	10872.920			
3	9224.4980	10837.7242			

INFRARED SPECTRUM OF ARGON

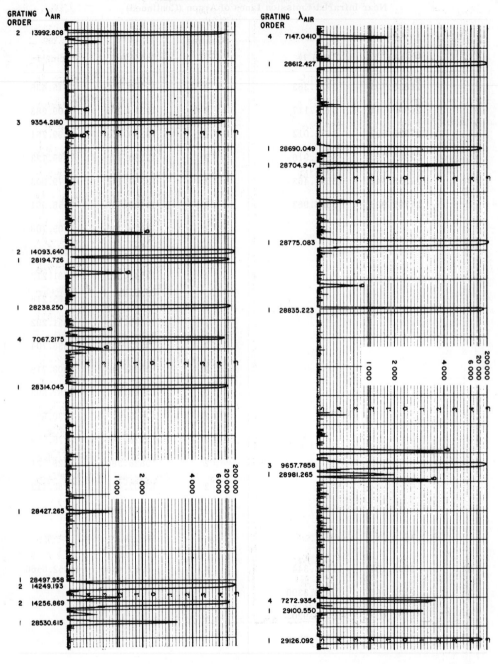

Grating order	λ_{air} (Å)	ν_{vac} (cm⁻¹)	Grating order	λ_{air} (Å)	ν_{vac} (cm⁻¹)
1	28194.726	3545.796	2	13992.808	7144.575
1	28238.250	3540.330	2	14093.640	7093.460
1	28314.045	3530.853	2	14249.193	7016.023
1	28427.265	3516.790	2	14256.869	7012.246
1	28497.958	3508.067			
1	28530.615	3504.051			
1	28612.427	3494.032			
1	28690.049	3484.579			
1	28704.947	3482.770			
1	28775.083	3474.281			
1	28835.223	3467.035			
1	28981.265	3449.564			
1	29100.550	3435.424			
1	29126.092	3432.412			
3	9354.2180	10687.4322	4	7067.2175	14145.9401
3	9657.7858	10351.5012	4	7147.0410	13987.9486
			4	7272.9354	13745.8188

INFRARED SPECTRUM OF ARGON

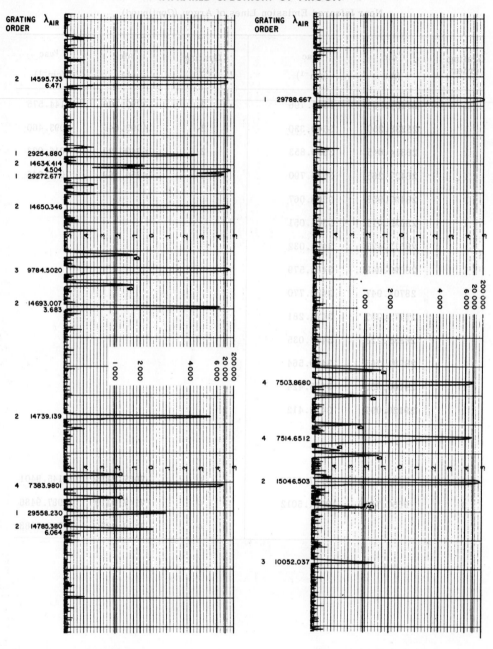

Wavelengths (Air) and Vacuum Wave Numbers (cm⁻¹) of the
Near Infrared Emission Lines of Argon (Continued)

Grating order	λ_{air} (Å)	ν_{vac} (cm⁻¹)	Grating order	λ_{air} (Å)	ν_{vac} (cm⁻¹)
1	29254.880	3417.301	2	14595.733	6849.445
1	29272.677	3415.223	2	14596.471	6849.099
1	29558.230	3382.230	2	14634.414	6831.341
1	29788.667	3356.066	2	14634.504	6831.299
			2	14650.346	6823.912
			2	14693.007	6804.099
			2	14693.683	6803.786
			2	14739.139	6782.803
			2	14785.380	6761.590
			2	14786.064	6761.277
			2	15046.503	6644.246
3	9784.5020	10217.4424	4	7383.9801	13539.1023
3	10052.037	9945.506	4	7503.8680	13322.7918
			4	7514.6512	13303.6742

INFRARED SPECTRUM OF ARGON

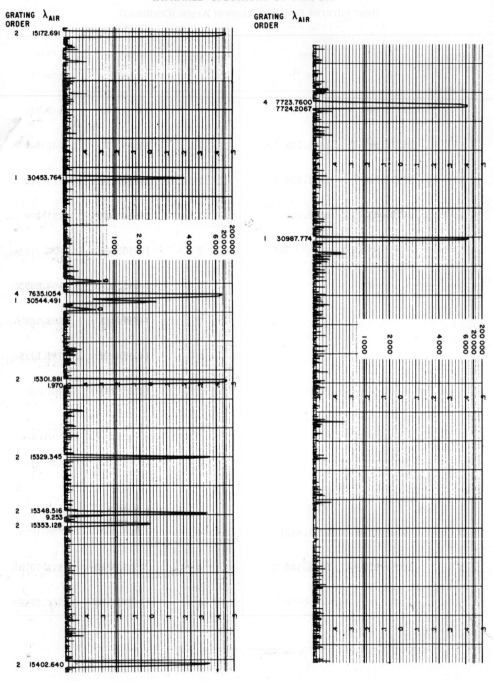

36

Wavelengths (Air) and Vacuum Wave Numbers (cm⁻¹) of the
Near Infrared Emission Lines of Argon (Continued)

Grating order	λ_{air} (Å)	ν_{vac} (cm⁻¹)	Grating order	λ_{air} (Å)	ν_{vac} (cm⁻¹)
1	30453.764	3282.771	2	15172.691	6588.988
1	30544.491	3273.020	2	15301.881	6533.359
1	30987.774	3226.199	2	15301.970	6533.321
			2	15329.345	6521.654
			2	15348.516	6513.508
			2	15349.253	6513.195
			2	15353.128	6511.551
			2	15402.640	6490.620
			4	7635.1054	13093.7918
			4	7723.7600	12943.4998
			4	7724.2067	12942.7514

INFRARED SPECTRUM OF ARGON

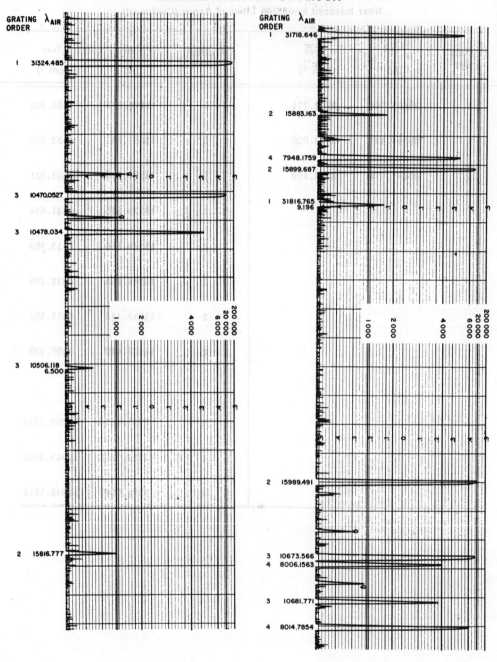

Wavelengths (Air) and Vacuum Wave Numbers (cm⁻¹) of the
Near Infrared Emission Lines of Argon (Continued)

Grating order	λ_{air} (Å)	ν_{vac} (cm⁻¹)	Grating order	λ_{air} (Å)	ν_{vac} (cm⁻¹)
1	31324.485	3191.521	2	15816.777	6320.674
1	31718.646	3151.860	2	15883.163	6294.255
1	31816.765	3142.140	2	15899.687	6287.714
1	31819.196	3141.900	2	15989.491	6252.399
3	10470.0527	9548.4338	4	7948.1759	12578.0434
3	10478.034	9541.161	4	8006.1563	12486.9540
3	10506.118	9515.656	4	8014.7854	12473.5100
3	10506.500	9515.310			
3	10673.566	9366.374			
3	10681.771	9359.179			

INFRARED SPECTRUM OF ARGON

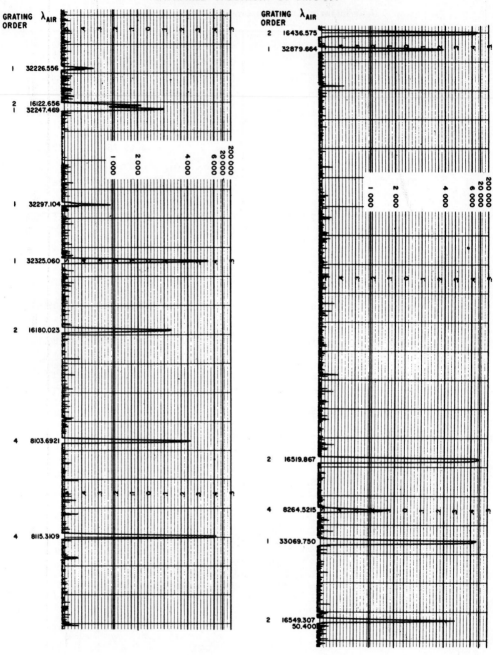

Wavelengths (Air) and Vacuum Wave Numbers (cm⁻¹) of the
Near Infrared Emission Lines of Argon (Continued)

Grating order	λ_{air} (Å)	ν_{vac} (cm⁻¹)	Grating order	λ_{air} (Å)	ν_{vac} (cm⁻¹)
1	32226.556	3102.185	2	16122.656	6200.758
1	32247.469	3100.173	2	16180.023	6178.773
1	32297.104	3095.408	2	16436.575	6082.331
1	32325.060	3092.732	2	16519.867	6051.664
1	32879.664	3040.565	2	16549.307	6040.899
1	33069.750	3023.087	2	16550.400	6040.500
			4	8103.6921	12336.6620
			4	8115.3109	12318.9996
			4	8264.5215	12096.5890

INFRARED SPECTRUM OF ARGON

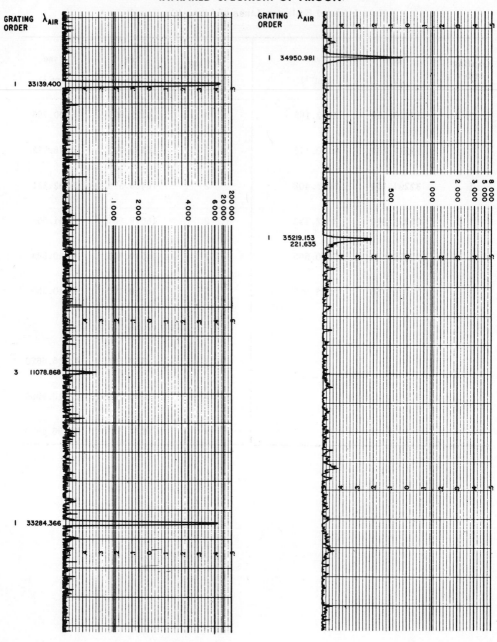

Wavelengths (Air) and Vacuum Wave Numbers (cm⁻¹) of the Near
Infrared Emission Lines of Argon (Continued)

Grating order	λ_{air} (Å)	ν_{vac} (cm⁻¹)
1	33139.400	3016.734
1	33284.366	3003.595
1	34950.981	2860.370
1	35219.153	2838.590
1	35221.635	2838.390
3	11078.868	9023.722

INFRARED SPECTRUM OF ARGON

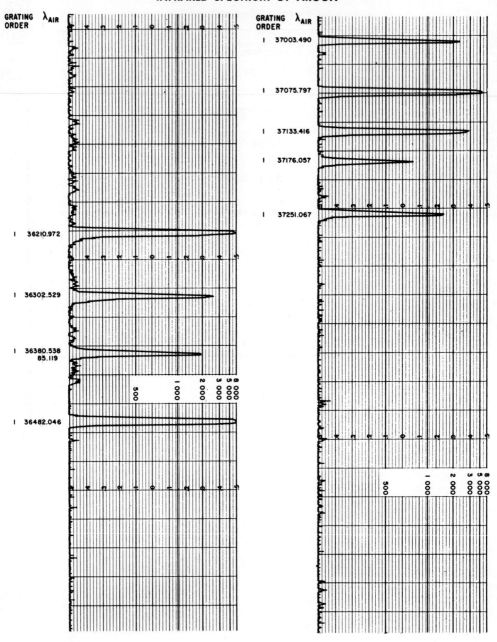

Wavelengths (Air) and Vacuum Wave Numbers (cm⁻¹) of the Near
Infrared Emission Lines of Argon (Continued)

Grating order	λ_{air} (Å)	ν_{vac} (cm⁻¹)
1	36210.972	2760.841
1	36302.529	2753.878
1	36380.538	2747.973
1	36385.119	2747.627
1	36482.046	2740.327
1	37003.490	2701.711
1	37075.797	2696.442
1	37133.416	2692.258
1	37176.057	2689.170
1	37251.067	2683.755

INFRARED SPECTRUM OF ARGON

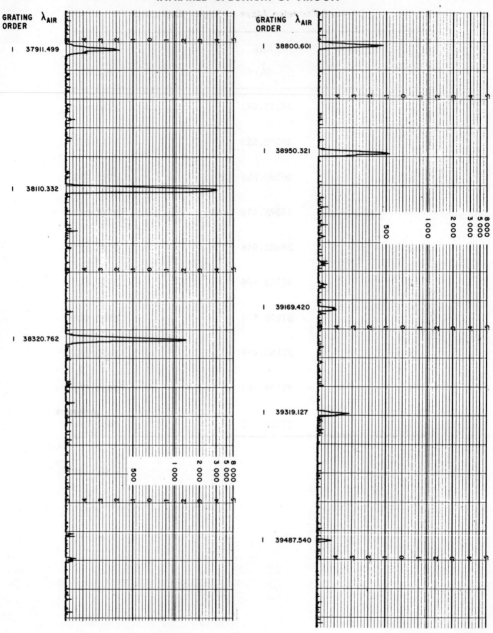

Wavelengths (Air) and Vacuum Wave Numbers (cm⁻¹) of the Near
Infrared Emission Lines of Argon (Continued)

Grating order	λ_{air} (Å)	ν_{vac} (cm⁻¹)
1	37911.499	2637.003
1	38110.332	2623.245
1	38320.762	2608.840
1	38800.601	2576.577
1	38950.321	2566.673
1	39169.420	2552.316
1	39319.127	2542.598
1	39487.540	2531.754

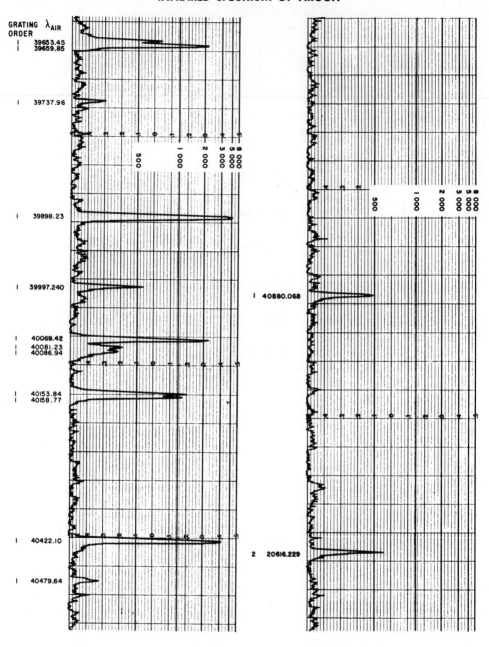

Wavelengths (Air) and Vacuum Wave Numbers (cm⁻¹) of the Near
Infrared Emission Lines of Argon (Continued)[a]

Grating order	λ_{air} (Å)	ν_{vac} (cm⁻¹)
1	39653.45	2521.16
1	39659.85	2520.75
1	39737.96	2515.80
1	39898.23	2505.69
1	39997.240	2499.491
1	40069.42	2494.99
1	40081.23	2494.25
1	40086.94	2493.90
1	40153.84	2489.74
1	40158.77	2489.44
1	40422.10	2473.22
1	40479.64	2469.70
1	40880.068	2445.513
2	20616.229	4849.224

[a] The argon spectrum in the region 3.9-4.1 microns is included here in order to illustrate the appearance of the spectrum in this region. Lines for which the one-hundredth place in the wave number is shown are transitions between levels not determined with sufficient precision until now. The three lines with λ39997.240, λ40880.068, and the second order λ20616.229 for which the one-thousandth place in the wave number is included may be considered as standards.

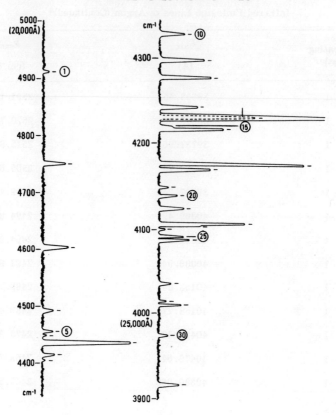

Wavelengths (Air) and Vacuum Wave Numbers (cm⁻¹) of the Near Infrared Emission Lines of Neon

Serial number (as in figure on opposite page)	λ_{air} (Å)	ν_{vac} (cm⁻¹)	Serial number (as in figure on opposite page)	λ_{air} (Å)	ν_{vac} (cm⁻¹)
1	20350.2377	4912.6065	17	23951.4175	4173.9793
2	21041.2948	4751.2624	18	23978.115	4169.332
3	21708.1449	4605.3090	19	24098.557	4148.494
4	22247.345	4493.692	20	24161.429	4137.699
5	22428.130	4457.470	21	24249.6384	4122.6479
6	22466.804	4449.797	22	24365.0477	4103.1203
7	22530.4039	4437.2359	23	24371.601	4102.017
8	22661.8132	4411.5057	24	24383.359	4100.039
9	22687.768	4406.459	25	24447.853	4089.223
10	23100.5137	4327.7271	26	24459.367	4087.298
11	23260.3018	4297.9975		24459.678	4087.246
12	23373.0002	4277.2737	27	24776.473	4034.986
13	23565.3623	4242.3587	28	24903.733	4014.367
14	23636.5149	4229.5880	29	24928.870	4010.319
15	23701.636	4217.967	30	25161.682	3973.213
16	23707.611	4216.904	31	25524.3660	3916.7564
	23709.1600	4216.6285			

INFRARED SPECTRUM OF KRYPTON

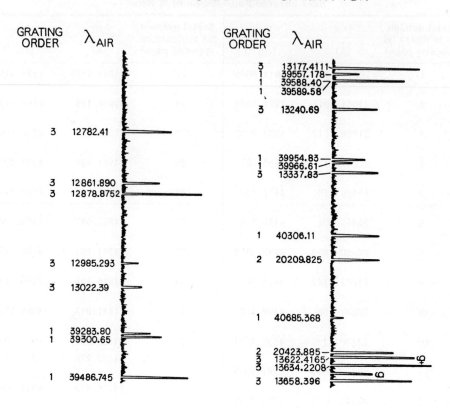

GRATING ORDER	λ_{AIR}
3	12782.41
3	12861.890
3	12878.8752
3	12985.293
3	13022.39
1	39283.80
1	39300.65
1	39486.745

GRATING ORDER	λ_{AIR}
3	13177.4111
1	39557.178
1	39588.40
1	39589.58
3	13240.69
1	39954.83
1	39966.61
3	13337.83
1	40306.11
2	20209.825
1	40685.368
2	20423.885
3	13622.4165
3	13634.2208
3	13658.396

Wavelengths (Air) and Vacuum Wave Numbers (cm⁻¹) of the Near Infrared Emission Lines of Krypton

Grating order	λ air (Å)	ν vac (cm⁻¹)	Int.	Grating order	λ air (Å)	ν vac (cm⁻¹)	Int.
1	39283.80	2544.88	50	1	39557.178	2527.297	40
1	39300.65	2543.79	90	1	39588.40	2525.30	500
1	39486.745	2531.805	140	1	39589.58	2525.23	
				1	39954.83	2502.14	60
3	12782.41	7821.11	150	1	39966.61	2501.41	30
3	12861.890	7772.781	80	1	40306.11	2480.34	130
3	12878.8752	7762.5300	750	1	40685.368	2457.216	20
3	12985.293	7698.914	25	2	20209.825	4946.738	140
3	13022.39	7676.98	30	2	20423.885	4894.892	300
				3	13177.4111	7586.6692	1100
					13177.4119	7586.6687 (Kr⁸⁶)	
				3	13240.69	7550.41	110
				3	13337.83	7495.42	120
				3	13622.4165	7338.8348	1000
					13622.4156	7338.8353 (Kr⁸⁶)	
				3	13634.2208	7332.4810	2400
					13634.2209	7332.4809 (Kr⁸⁶)	
				3	13658.396	7319.502₅	800

INFRARED SPECTRUM OF KRYPTON

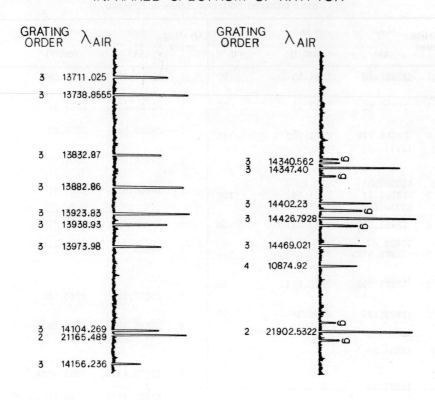

GRATING ORDER	λ_AIR		GRATING ORDER	λ_AIR
3	13711.025			
3	13738.8555			
3	13832.87		3	14340.562
			3	14347.40
3	13882.86			
			3	14402.23
3	13923.83		3	14426.7928
3	13938.93			
3	13973.98		3	14469.021
			4	10874.92
3	14104.269		2	21902.5322
2	21165.489			
3	14156.236			

Wavelengths (Air) and Vacuum Wave Numbers (cm⁻¹) of the
Near Infrared Emission Lines of Krypton (Continued)

Grating order	λ_{air} (Å)	ν_{vac} (cm⁻¹)	Int.	Grating order	λ_{air} (Å)	ν_{vac} (cm⁻¹)	Int.
2	21165.489	4723.383	600	2	21902.5322	4564.4364	1800
					21902.5111	4564.4408(Kr^{86})	
3	13711.025	7291.407	200				
				3	14340.562	6971.322	30
3	13738.8555	7276.6371	600				
	13738.8500	7276.6400(Kr^{86})		3	14347.40	6968.00	800
3	13832.87	7227.18	150	3	14402.23	6941.47	180
3	13882.86	7201.16	500	3	14426.7928	6929.6534	2000
					14426.7935	6929.6531(Kr^{86})	
3	13923.83	7179.97	700				
				3	14469.021	6909.429	140
3	13938.93	7172.19	200				
3	13973.98	7154.20	150				
				4	10874.92	9192.95	80
3	14104.269	7088.114	140				
3	14156.236	7062.094	50				

INFRARED SPECTRUM OF KRYPTON

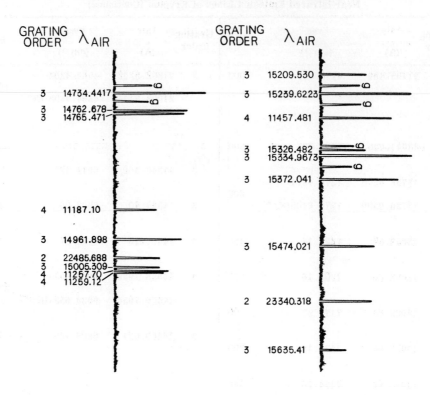

Wavelengths (Air) and Vacuum Wave Numbers (cm⁻¹) of the
Near Infrared Emission Lines of Krypton (Continued)

Grating order	λ_{air} (Å)	ν_{vac} (cm⁻¹)	Int.	Grating order	λ_{air} (Å)	ν_{vac} (cm⁻¹)	Int.
2	22485.688	4446.060	120	2	23340.318	4283.263	180
3	14734.4417	6784.9654	1600	3	15209.530	6573.029	140
	14734.4360	6784.9680(Kr⁸⁶)		3	15239.6223	6560.0498	1700
3	14762.678	6771.988	550		15239.6159	6560.0525(Kr⁸⁶)	
3	14765.471	6770.707	450	3	15326.482	6522.872	130
3	14961.898	6681.818	400	3	15334.9673	6519.2628	1500
3	15005.309	6662.487	120		15334.9611	6519.2654(Kr⁸⁶)	
				3	15372.041	6503.540	700
4	11187.10	8936.42	100	3	15474.021	6460.679	200
4	11257.70	8880.38	200	3	15635.41	6393.99	40
4	11259.12	8879.26	150				
				4	11457.481	8725.533	500

INFRARED SPECTRUM OF KRYPTON

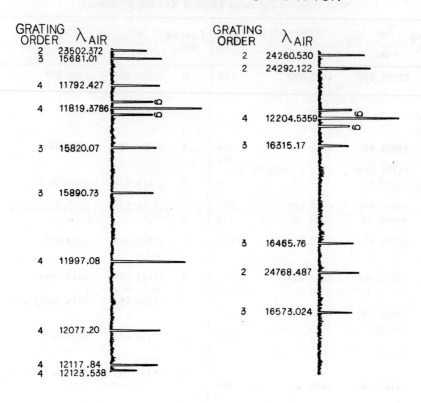

Wavelengths (Air) and Vacuum Wave Numbers (cm⁻¹) of the Near Infrared Emission Lines of Krypton (Continued)

Grating order	λ_{air} (Å)	ν_{vac} (cm⁻¹)	Int.	Grating order	λ_{air} (Å)	ν_{vac} (cm⁻¹)	Int.
2	23502.372	4253.729	70	2	24260.530	4120.797	120
				2	24292.122	4115.438	180
3	15681.01	6375.40	180	2	24768.487	4036.287	90
3	15820.07	6319.36	120				
3	15890.73	6291.26	100	3	16315.17	6127.59	50
				3	16465.76	6071.55	70
4	11792.427	8477.698	150	3	16573.024	6032.254	70
4	11819.3786	8458.3666	1500				
	11819.3759	8458.3685 (Kr⁸⁶)		4	12204.5359	8191.4335	800
4	11997.08	8333.08	600				
4	12077.20	8277.80	160				
4	12117.84	8250.04	140				
4	12123.538	8246.161	40				

INFRARED SPECTRUM OF KRYPTON

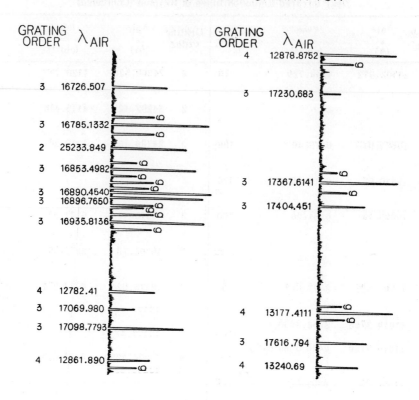

GRATING ORDER λ_{AIR}

3	16726.507
3	16785.1332
2	25233.849
3	16853.4982
3	16890.4540
3	16896.7650
3	16935.8136
4	12782.41
3	17069.980
3	17098.7793
4	12861.890

GRATING ORDER λ_{AIR}

4	12878.8752
3	17230.683
3	17367.6141
3	17404.451
4	13177.4111
3	17616.794
4	13240.69

Grating order	λ air (Å)	ν vac (cm⁻¹)	Int.	Grating order	λ air (Å)	ν vac (cm⁻¹)	Int.
2	25233.849	3961.850	600	3	17230.683	5802.016	30
				3	17367.6141	5856.2712	700
					17367.6050	5856.2742(Kr⁸⁶)	
3	16726.507	5976.902	200				
				3	17404.451	5744.088	120
3	16785.1332	5956.0262	2000				
	16785.1275	5956.0282(Kr⁸⁶)		3	17616.794	5674.852	150
3	16853.4982	5931.8660	1000				
	16853.4885	5931.8694(Kr⁸⁶)					
				4	12878.8752	7762.5300	600
3	16890.4540₅	5918.8873	2400				
	16890.4409	5918.8919(Kr⁸⁶)		4	13177.4111	7586.6692	600
					13177.4119	7586.6687(Kr⁸⁶)	
3	16896.7650	5916.6766	1600				
	16896.7530	5966.6808(Kr⁸⁶)		4	13240.69	7550.41	90
3	16935.8136	5903.0346	1800				
	16935.8057	5903.0374(Kr⁸⁶)					
3	17069.980	5856.638	40				
3	17098.7793	5846.7738	600				
	17098.7696	5846.7771(Kr⁸⁶)					
4	12782.41	7821.11	120				
4	12861.890	7772.781	100				

INFRARED SPECTRUM OF KRYPTON

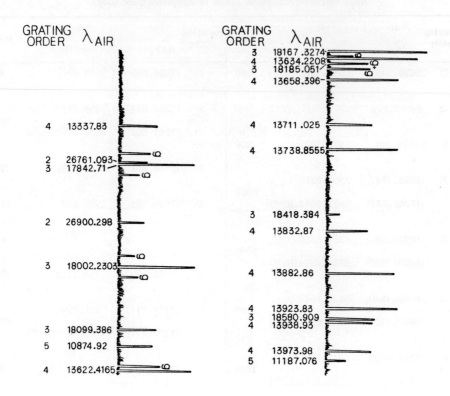

GRATING ORDER	λ_AIR
4	13337.83
2	26761.093
3	17842.71
2	26900.298
3	18002.2303
3	18099.386
5	10874.92
4	13622.4165

GRATING ORDER	λ_AIR
3	18167.3274
4	13634.2208
3	18185.051
4	13658.396
4	13711.025
4	13738.8555
3	18418.384
4	13832.87
4	13882.86
4	13923.83
3	18580.909
4	13938.93
4	13973.98
5	11187.076

Wavelengths (Air) and Vacuum Wave Numbers (cm⁻¹) of the Near Infrared Emission Lines of Krypton (Continued)

Grating order	λ air (Å)	ν vac (cm⁻¹)	Int.	Grating order	λ air (Å)	ν vac (cm⁻¹)	Int.
2	26761.093	3735.749	50	3	18167.3274₅	5502.8843	2600
					18167.3153	5502.8880(Kr⁸⁶)	
2	26900.298	3716.417	40				
				3	18185.051	5497.521	90
				3	18418.384	5427.876	20
3	17842.71	5603.00	650				
	17842.7376	5602.9910(Kr⁸⁶)		3	18580.909	5380.399	150
3	18002.2303	5553.3508	700				
	18002.2291	5553.3512(Kr⁸⁶)		4	13634.2208	7332.4810	1500
3	18099.386	5523.541	80		13634.2209	7332.4809(Kr⁸⁶)	
				4	13658.396	7319.502₅	500
4	13337.83	7495.42	80	4	13711.025	7291.407	110
4	13622.4165	7338.8348	600	4	13738.8555	7276.6371	500
	13622.4156	7338.8353(Kr⁸⁶)			13738.8500	7276.6400(Kr⁸⁶)	
				4	13832.87	7227.18	90
5	10874.92	9192.95	70	4	13882.86	7201.16	400
				4	13923.83	7179.97	500
				4	13938.93	7172.19	130
				4	13973.98	7154.20	110
				5	11187.076	8936.439	30

INFRARED SPECTRUM OF KRYPTON

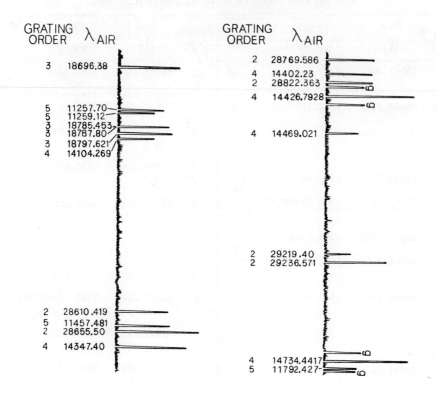

GRATING ORDER	λ_AIR
3	18696.38
5	11257.70
5	11259.12
3	18785.453
3	18787.80
3	18797.621
4	14104.269
2	28610.419
5	11457.481
2	28655.50
4	14347.40

GRATING ORDER	λ_AIR
2	28769.586
4	14402.23
2	28822.363
4	14426.7928
4	14469.021
2	29219.40
2	29236.571
4	14734.4417
5	11792.427

Wavelengths (Air) and Vacuum Wave Numbers (cm⁻¹) of the
Near Infrared Emission Lines of Krypton (Continued)

Grating order	λ_{air} (Å)	ν_{vac} (cm⁻¹)	Int.	Grating order	λ_{air} (Å)	ν_{vac} (cm⁻¹)	Int.
2	28610.419	3494.277	180	2	28769.586	3474.945	150
2	28655.50	3488.78	1000	2	28822.363	3468.582	140
	28655.7172	3488.7533(Kr⁸⁶)		2	29219.40	3421.45	40
				2	29236.571	3419.441	300
3	18696.38	5347.17	300				
3	18785.453	5321.815	170	4	14402.23	6941.47	130
3	18787.80	5321.15	50	4	14426.7928	6929.6534	1200
3	18797.621	5318.370	200		14426.7935	6929.6531(Kr⁸⁶)	
				4	14469.021	6909.429	60
4	14104.269	7088.114	70	4	14734.4417	6784.9654	1000
4	14347.40	6968.00	500		14734.4360	6784.9680(Kr⁸⁶)	
5	11257.70	8880.38	120	5	11792.427	8477.698	70
5	11259.12	8879.26	70				
5	11457.481	8725.533	200				

INFRARED SPECTRUM OF KRYPTON

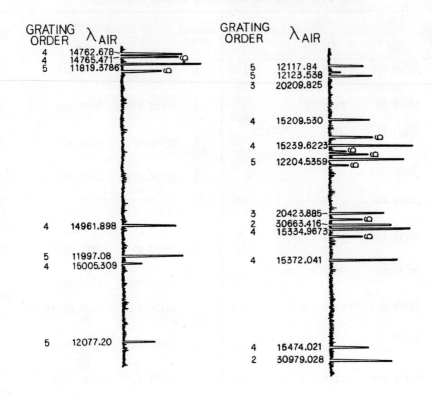

GRATING ORDER	λ AIR
4	14762.678
4	14765.471
5	11819.3786
4	14961.898
5	11997.08
4	15005.309
5	12077.20

GRATING ORDER	λ AIR
5	12117.84
5	12123.538
3	20209.825
4	15209.530
4	15239.6223
5	12204.5359
3	20423.885
2	30663.416
4	15334.9673
4	15372.041
4	15474.021
2	30979.028

Grating order	λ air (Å)	ν vac (cm⁻¹)	Int.	Grating order	λ air (Å)	ν vac (cm⁻¹)	Int.
4	14762.678	6771.988	300	2	30663.416	3260.326	300
4	14765.471	6770.707	250	2	30979.028	3227.110	300
4	14961.898	6681.818	200				
4	15005.309	6662.487	30	3	20209.825	4946.738	110
				3	20423.885	4894.892	200
5	11819.3786	8458.3666	800				
	11819.3759	8458.3685(Kr⁸⁶)		4	15209.530	6573.029	90
5	11997.08	8333.08	300	4	15239.6223	6560.0498	1000
5	12077.20	8277.80	60		15239.6159	6560.0525(Kr⁸⁶)	
				4	15326.482	6522.872	40
				4	15334.9673	6519.2628	800
					15334.9611	6519.2654(Kr⁸⁶)	
				4	15372.041	6503.540	400
				4	15474.021	6460.679	80
				5	12117.84	8250.04	70
				5	12123.538	8246.161	20
				5	12204.5359	8191.4335	600

INFRARED SPECTRUM OF XENON

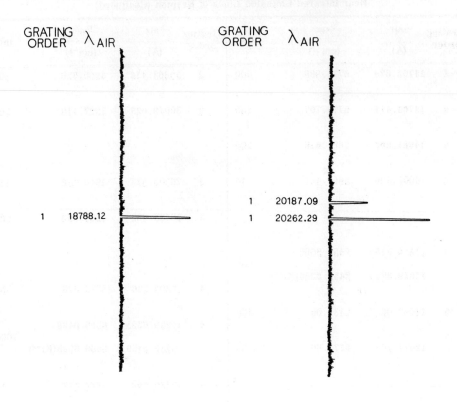

GRATING ORDER	λ_{AIR}		GRATING ORDER	λ_{AIR}
1	18788.12		1	20187.09
			1	20262.29

Wavelengths (Air) and Vacuum Wave Numbers (cm⁻¹) of the
Near Infrared Emission Lines of Xenon

Grating order	λ air (Å)	ν vac (cm⁻¹)	Int.	Grating order	λ air (Å)	ν vac (cm⁻¹)	Int.
1	18788.12	5321.06	500	1	20187.09	4952.31	80
				1	20262.29	4933.93	2500
					20262.2395	4933.9418 (Xe¹³⁶)	

INFRARED SPECTRUM OF XENON

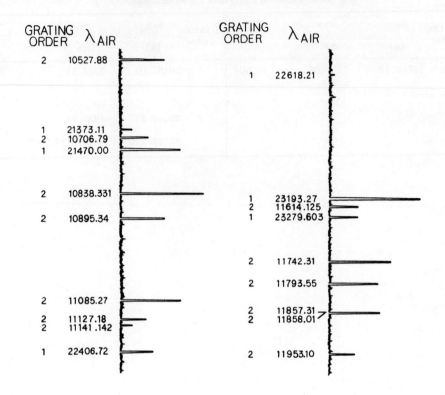

GRATING
ORDER λ_{AIR}

2	10527.88
1	21373.11
2	10706.79
1	21470.00
2	10838.331
2	10895.34
2	11085.27
2	11127.18
2	11141.142
1	22406.72

GRATING
ORDER λ_{AIR}

1	22618.21
1	23193.27
2	11614.125
1	23279.603
2	11742.31
2	11793.55
2	11857.31
2	11858.01
2	11953.10

Grating order	λ_{air} (Å)	ν_{vac} (cm⁻¹)	Int.	Grating order	λ_{air} (Å)	ν_{vac} (cm⁻¹)	Int.
1	21373.11	4677.50	20	1	22618.21	4420.01	10
1	21470.00	4656.39	300	1	23193.27	4310.42	1500
					23193.3328	4310.4076(Xe¹³⁶)	
1	22406.72	4461.73	60	1	23279.603	4294.434	50
				2	11614.125	8607.849	50
2	10527.88	9495.99	120				
				2	11742.31	8513.88	300
2	10706.79	9337.31	50				
				2	11793.55	8476.89	150
2	10838.331	9223.986	1000				
				2	11857.31	8431.31	160
2	10895.34	9175.72	120		11858.01	8430.81	
2	11085.27	9018.51	300	2	11953.10	8363.74	40
2	11127.18	8984.54	40				
2	11141.142	8973.283	20				

INFRARED SPECTRUM OF XENON

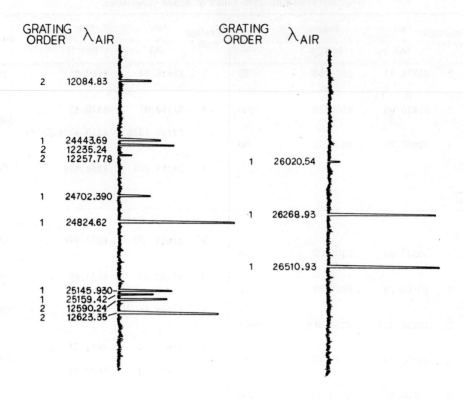

Wavelengths (Air) and Vacuum Wave Numbers (cm⁻¹) of the
Near Infrared Emission Lines of Xenon (Continued)

Grating order	λ_{air} (Å)	ν_{vac} (cm⁻¹)	Int.	Grating order	λ_{air} (Å)	ν_{vac} (cm⁻¹)	Int.
1	24443.69	4089.92	100	1	26020.54	3842.07	20
1	24702.390	4047.087	60	1	26268.93	3805.74	3500
					26269.0832	3805.7181(Xe¹³⁶)	
1	24824.62	4027.16	5500	1	26510.93	3771.00	4500
	24824.7157	4027.1447(Xe¹³⁶)			26510.8645	3771.0096(Xe¹³⁶)	
1	25145.930	3975.702	200				
1	25159.42	3973.57	70				
2	12084.83	8272.57	60				
2	12235.24	8170.88	200				
2	12257.778	8155.854	20				
2	12590.24	7940.49	150				
2	12623.35	7919.66	2500				

INFRARED SPECTRUM OF XENON

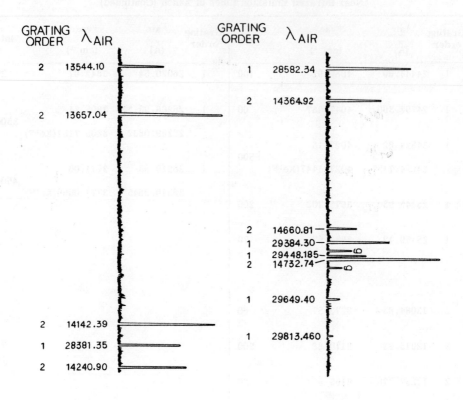

GRATING ORDER	λ_{AIR}
2	13544.10
2	13657.04
2	14142.39
1	28381.35
2	14240.90

GRATING ORDER	λ_{AIR}
1	28582.34
2	14364.92
2	14660.81
1	29384.30
1	29448.185
2	14732.74
1	29649.40
1	29813.460

Wavelengths (Air) and Vacuum Wave Numbers (cm⁻¹) of the Near Infrared Emission Lines of Xenon (Continued)

Grating order	λ_{air} (Å)	ν_{vac} (cm⁻¹)	Int.	Grating order	λ_{air} (Å)	ν_{vac} (cm⁻¹)	Int.
1	28381.35	3522.48	300	1	28582.34	3497.71	1000
				1	29384.30	3402.25	300
2	13544.10	7381.27	120	1	29448.185	3394.869	80
2	13657.04	7320.23	3000	1	29649.40	3371.83	20
2	14142.39	7069.01	2000	1	29813.460	3353.275	10
2	14240.90	7020.11	600				
				2	14364.92	6959.50	200
				2	14660.81	6819.04	50
				2	14732.74	6785.75	5000

INFRARED SPECTRUM OF XENON

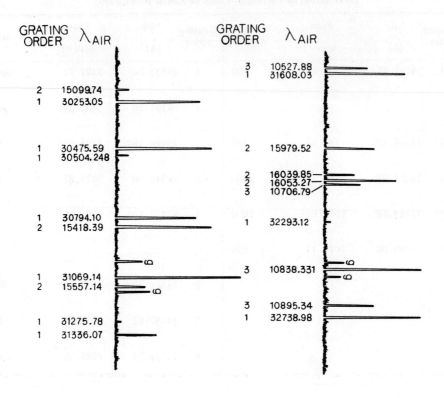

GRATING ORDER	λ_AIR
2	15099.74
1	30253.05
1	30475.59
1	30504.248
1	30794.10
2	15418.39
1	31069.14
2	15557.14
1	31275.78
1	31336.07

GRATING ORDER	λ_AIR
3	10527.88
1	31608.03
2	15979.52
2	16039.85
2	16053.27
3	10706.79
1	32293.12
3	10838.331
3	10895.34
1	32738.98

Wavelengths (Air) and Vacuum Wave Numbers (cm⁻¹) of the
·Near Infrared Emission Lines of Xenon (Continued)

Grating order	λ air (Å)	ν vac (cm⁻¹)	Int.	Grating order	λ air (Å)	ν vac (cm⁻¹)	Int.
1	30253.05	3304.55	1000	1	31608.03	3162.89	800
1	30475.59	3280.42	2000	1	32293.12	3095.79	10
	30475.4527	3280.4347 (Xe^{136})					
				1	32738.98	3053.63	2000
1	30504.248	3277.338	20		32739.2788	3053.6022 (Xe^{136})	
1	30794.10	3246.49	800				
1	31069.14	3217.75	10000	2	15979.52	6256.30	150
	31069.2302	3217.7409 (Xe^{136})		2	16039.85	6232.77	50
1	31275.78	3196.49	10	2	16053.27	6227.56	500
1	31336.07	3190.34	90				
				3	10527.88	9495.99	100
2	15099.74	6620.82	20	3	10706.79	9337.31	70
2	15418.39	6483.99	2000	3	10838.331	9223.986	2000
2	15557.14	6426.16	50	3	10895.34	9175.72	150

INFRARED SPECTRUM OF XENON

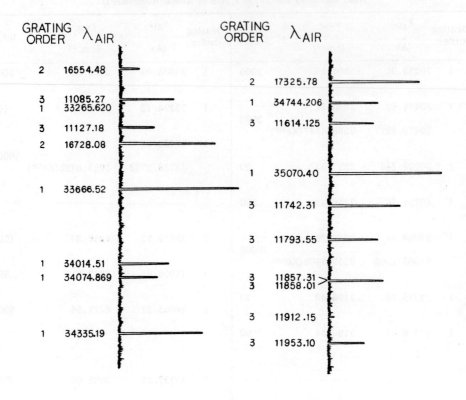

Grating order	λ_{air} (Å)	ν_{vac} (cm⁻¹)	Int.	Grating order	λ_{air} (Å)	ν_{vac} (cm⁻¹)	Int.
1	33265.620	3005.287	10	1	34744.206	2877.393	150
1	33666.52	2969.50	7000	1	35070.40	2850.63	5000
	33666.6991	2969.4843 (Xe¹³⁶)			35070.2520	2850.6420 (Xe¹³⁶)	
1	34014.51	2939.12	150				
1	34074.869	2933.914	80	2	17325.78	5770.17	1500
1	34335.19	2911.67	1000				
				3	11614.125	8607.849	120
2	16554.48	6039.01	30	3	11742.31	8513.88	500
2	16728.08	5976.34	2000	3	11793.55	8476.89	150
				3	11857.31	8431.31 ⎫	200
3	11085.27	9018.51	200		11858.01	8430.81 ⎭	
3	11127.18	8984.54	70	3	11912.15	8392.49	10
				3	11953.10	8363.74	70

INFRARED SPECTRUM OF XENON

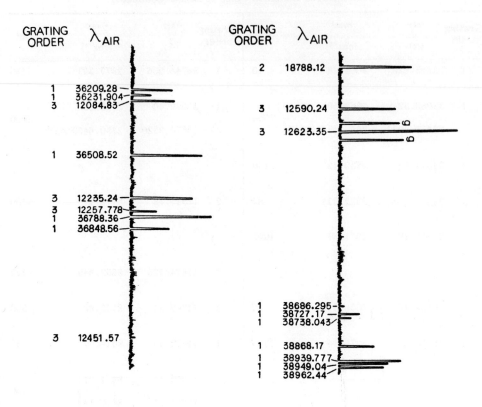

GRATING ORDER | λ_{AIR}
1   36209.28
1   36231.904
3   12084.83
1   36508.52
3   12235.24
3   12257.778
1   36788.36
1   36848.56
3   12451.57

GRATING ORDER | λ_{AIR}
2   18788.12
3   12590.24
3   12623.35
1   38686.295
1   38727.17
1   38738.043
1   38868.17
1   38939.777
1   38949.04
1   38962.44

Wavelengths (Air) and Vacuum Wave Numbers (cm⁻¹) of the Near Infrared Emission Lines of Xenon (Continued)

Grating order	λ_{air} (Å)	ν_{vac} (cm⁻¹)	Int.	Grating order	λ_{air} (Å)	ν_{vac} (cm⁻¹)	Int.
1	36209.28	2760.97	90	1	38686.295	2584.190	10
1	36231.904	2759.246	30	1	38727.17	2581.46	30
1	36508.52	2738.34	500	1	38738.043	2580.738	20
1	36788.36	2717.51	800	1	38868.17	2572.10	70
1	36848.56	2713.07	80	1	38939.777	2567.368	300
				1	38949.04	2566.76	150
3	12084.83	8272.57	100	1	38962.44	2565.87	120
3	12235.24	8170.88	300				
3	12257.778	8155.854	40	2	18788.12	5321.06	500
3	12451.57	8028.92	10				
				3	12590.24	7940.49	200
				3	12623.35	7919.66	6000

INFRARED SPECTRUM OF XENON

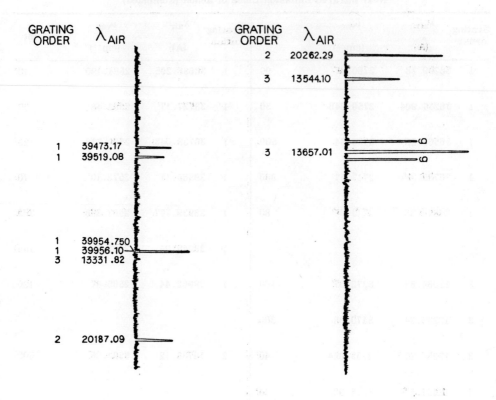

GRATING ORDER	λ_{AIR}
1	39473.17
1	39519.08
1	39954.750
1	39956.10
3	13331.82
2	20187.09

GRATING ORDER	λ_{AIR}
2	20262.29
3	13544.10
3	13657.01

Wavelengths (Air) and Vacuum Wave Numbers (cm⁻¹) of the
Near Infrared Emission Lines of Xenon (Continued)

Grating order	λ_{air} (Å)	ν_{vac} (cm⁻¹)	Int.	Grating order	λ_{air} (Å)	ν_{vac} (cm⁻¹)	Int.
1	39473.17	2532.68	70	2	20262.29	4933.93	2000
1	39519.08	2529.73	50				
1	39954.750	2502.149	200	3	13544.10	7381.27	200
1	39956.10	2502.06		3	13657.01	7320.23	8000
2	20187.09	4952.31	80				
3	13331.82	7498.80	10				

CHAPTER III

Molecular Absorption Standards in the Near Infrared

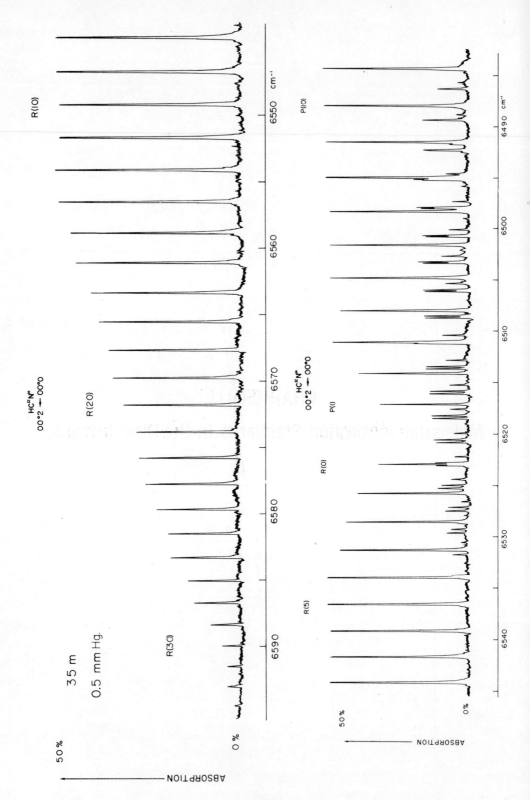

Vacuum Wave Numbers (cm⁻¹) for the Rotational Lines of the 00°2 ←— 00°0 (2ν₃)

Band of the HC¹²N¹⁴ Molecule

J	ν_{vac} cm⁻¹ (obs)	cm⁻¹ (calc)	J	ν_{vac} cm⁻¹ (obs)	cm⁻¹ (calc)	J	ν_{vac} cm⁻¹ (obs)	cm⁻¹ (calc)
R 32		6593.0916	R 17		6565.5351	R 2	6528.2296	6528.2297
31	6591.5597	91.5607	16		63.3496	1		25.4004
30	89.9866	89.9859	15		61.1207	0		22.5286
29	88.3667	88.3673	14		58.8487	P 1	16.6581	16.6581
28	86.7053	86.7048	13		56.5334	2	13.6580	13.6594
27		84.9986	12		54.1750	3		10.6186
26		83.2485	11		51.7735	4	07.5342	07.5357
25	81.4554	81.4548	10		49.3289	5	04.4087	04.4108
24	79.6181	79.6173	9		46.8413	6	01.2432	01.2440
23	77.7356	77.7362	8		44.3109	7	6498.0342	6498.0353
22	75.8107	75.8115	7	6541.7383	41.7375	8		94.7848
21	73.8427	73.8431	6	39.1201	39.1214	9		91.4925
20	71.8319	71.8313	5	36.4607	36.4624	10		88.1587
19	69.7759	69.7759	4	33.7608	33.7608	11		84.7833
18	67.6773	67.6773	3	31.0169	31.0165			

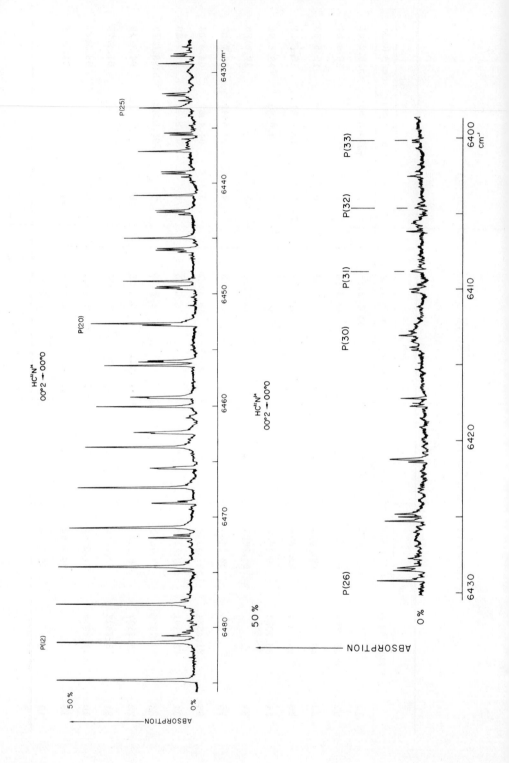

Vacuum Wave Numbers (cm⁻¹) for the Rotational Lines of the 00°2 ← 00°0 (2ν₃)

Band of the HC¹²N¹⁴ Molecule (Continued)

J	ν_{vac} cm⁻¹ (obs)	ν_{vac} cm⁻¹ (calc)	J	ν_{vac} cm⁻¹ (obs)	ν_{vac} cm⁻¹ (calc)
P 11		6484.7833	P 23	6441.0698	6441.0692
12		81.3665	24	37.1602	37.1616
13		77.9083	25	33.2133	33.2136
14		74.4088	26		29.2256
15		70.8680	27		25.1974
16		67.2862	28		21.1291
17		63.6633	29		17.0211
18		59.9995	30	12.8734	12.8733
19		56.2948	31	08.6857	08.6857
20		52.5493	32	04.4588	04.4586
21		48.7631	33	6400.1931	6400.1919
22	6444.9348	44.9364			

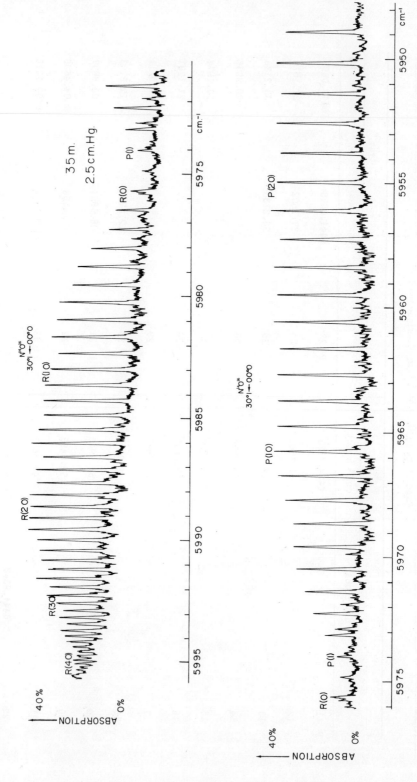

Vacuum Wave Numbers (cm⁻¹) for the Rotational Lines of the $30°1 \longleftarrow 00°0$ $(3\nu_1 + \nu_3)$ Band of the $N_2^{14}O^{16}$ Molecule

J	ν_{vac} cm⁻¹ (obs)	cm⁻¹ (calc)
R 40		5994.9172
39		94.7552
38		94.5772
37	5994.3834	94.3831
36	94.1717	94.1729
35	93.9438	93.9466
34	93.7083	93.7041
33	93.4463	93.4454
32	93.1729	93.1705
31	92.8789	92.8794
30	92.5723	92.5719
29	92.2470	92.2481
28	91.9059	91.9081
27	91.5498	91.5516
26	91.1816	91.1788
25	90.7918	90.7895
24	90.3856	90.3838
23	89.9627	89.9616
22	89.5249	89.5229
21	89.0688	89.0678
20	88.5954	88.5961
19	88.1087	88.1079

J	ν_{vac} cm⁻¹ (obs)	cm⁻¹ (calc)
R 18	5987.6006	5987.6031
17	87.0851	87.0817
16	86.5482	86.5438
15	85.9900	85.9892
14	85.4194	85.4180
13	84.8305	84.8302
12	84.2255	84.2258
11	83.6033	83.6046
10	82.9642	82.9669
9	82.3117	82.3124
8	81.6434	81.6414
7	80.9540	80.9536
6	80.2517	80.2491
5	79.5283	79.5279
4	78.7905	78.7901
3	78.0320	78.0355
2		77.2643
1		76.4763
0		75.6717
P 1	74.0068	74.0123
2	73.1565	73.1576
3	72.2830	72.2862

J	ν_{vac} cm⁻¹ (obs)	cm⁻¹ (calc)
P 4	5971.3989	5971.3982
5	70.4988	70.4935
6	69.5770	69.5721
7	68.6303	68.6340
8	67.6782	67.6794
9	66.7043	66.7081
10	65.7149	65.7202
11	64.7114	64.7156
12	63.6950	63.6945
13	62.6615	62.6569
14	61.6063	61.6026
15	60.5331	60.5319
16	59.4453	59.4446
17	58.3404	58.3408
18	57.2176	57.2205
19	56.0845	56.0838
20	54.9335	54.9306
21	53.7628	53.7611
22	52.5733	52.5751
23	51.3696	51.3728
24	50.1515	50.1542
25	48.9195	48.9192

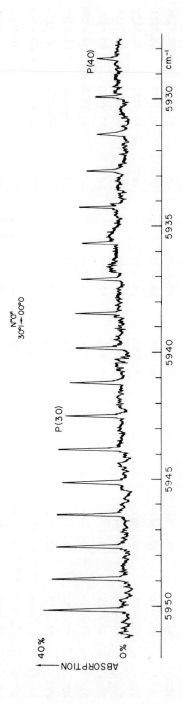

Vacuum Wave Numbers (cm^{-1}) for the Rotational Lines of the
30°1 ⟶ 00°0 ($3\nu_1 + \nu_3$) Band of the N$_2^{14}$O^{16} Molecule (Continued)

J	ν_{vac}	
	cm^{-1} (obs)	cm^{-1} (calc)
P 24	5950.1515	5950.1542
25	48.9195	48.9192
26	47.6685	47.6680
27	46.3974	46.4005
28	45.1110	45.1168
29	43.8204	43.8169
30	42.5017	42.5008
31	41.1679	41.1687
32	39.8178	39.8204
33	38.4540	38.4561
34	37.0762	37.0758
35	35.6797	35.6795
36	34.2661	34.2673
37	32.8400	32.8392
38	31.3976	31.3952
39	29.9344	29.9354
40		28.4598

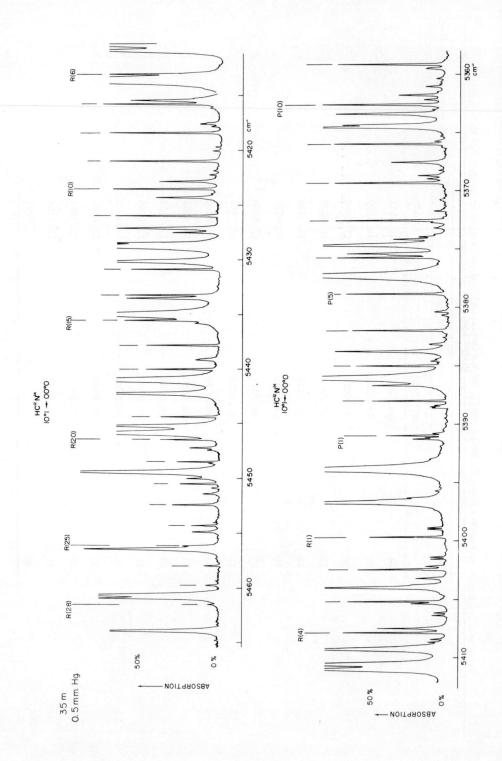

HC^{12}N^{14}
10°1 ← 00°0

35 m
0.5 mm. Hg

ABSORPTION

HC^{12}N^{14}
10°1 ← 00°0

ABSORPTION

94

Vacuum Wave Numbers (cm⁻¹) for the Rotational Lines of the $10°1 \longleftarrow 00°0$ $(\nu_1 + \nu_3)$ Band of the HC^{12}N^{14} Molecule

J	ν_{vac} cm⁻¹ (obs)	ν_{vac} cm⁻¹ (calc)	J	ν_{vac} cm⁻¹ (obs)	ν_{vac} cm⁻¹ (calc)	J	ν_{vac} cm⁻¹ (obs)	ν_{vac} cm⁻¹ (calc)
R 28	5459.762	5461.513	R 14	5433.137	5433.138	R 1	5399.486	5399.489
27		59.760	13	30.800	30.797	0		96.614
26	57.970	57.965	12		28.415	P 1	90.744	90.742
25		56.127	11	25.990	25.991	2	87.744	87.745
24	54.249	54.247	10	23.526	23.526	3		84.708
23		52.325	9	21.022	21.020	4		81.630
22	50.359	50.361	8	18.473	18.472	5		78.512
21	48.355	48.355	7	15.889	15.883	6		75.354
20	46.301	46.307	6		13.253	7	72.156	72.155
19	44.216	44.216	5		10.582	8	68.920	68.917
18		42.084	4	07.874	07.870	9		65.638
17	39.908	39.910	3	05.122	05.118	10		62.320
16	37.693	37.694	2	02.316	02.324	11		58.962
15		35.437						

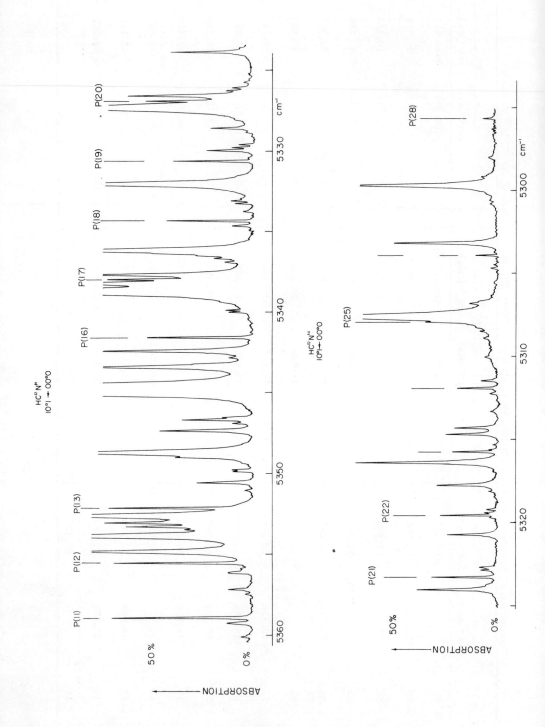

Vacuum Wave Numbers (cm⁻¹) for the Rotational Lines of the $10°1 \longleftarrow 00°0$ $(\nu_1 + \nu_3)$

Band of the HC^{12}N^{14} Molecule (Continued)

J	ν_{vac}	
	cm⁻¹ (obs)	cm⁻¹ (calc)
P 11	5355.563	5358.962
12		55.564
13		52.126
14		48.649
15		45.132
16		41.575
17		37.979
18	34.342	34.344
19	30.669	30.670

J	ν_{vac}	
	cm⁻¹ (obs)	cm⁻¹ (calc)
P 20		5326.956
21	5323.201	23.203
22	19.411	19.412
23	15.581	15.581
24	11.712	11.711
25		07.803
26	03.856	03.856
27		5299.870
28	5295.847	95.846

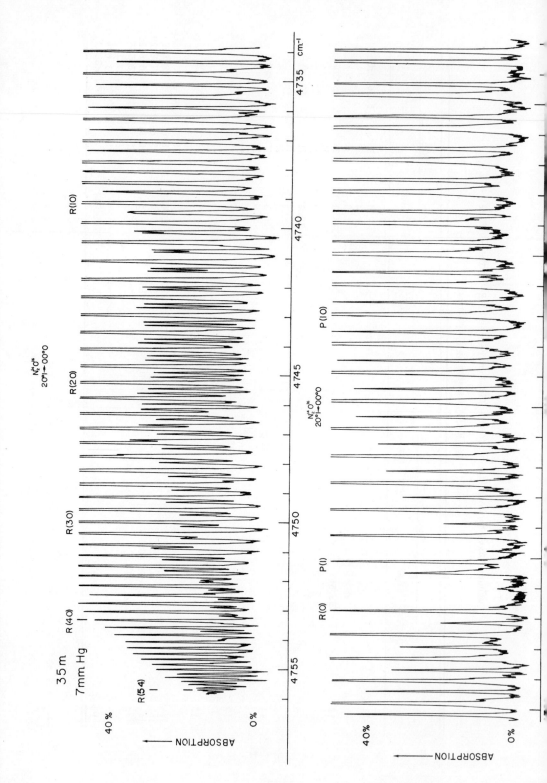

Vacuum Wave Numbers (cm^{-1}) for the Rotational Lines of the $20°1 \leftarrow 00°0$ $(2\nu_1 + \nu_3)$ Band of the $N_2^{14}O^{16}$ Molecule

J	ν_{vac} cm^{-1} (obs)	ν_{vac} cm^{-1} (calc)	J	ν_{vac} cm^{-1} (obs)	ν_{vac} cm^{-1} (calc)	J	ν_{vac} cm^{-1} (obs)	ν_{vac} cm^{-1} (calc)
R 54	4755.7236	4755.7227	R 29	4749.5542	4749.5518	R 5	4735.5646	4735.5659
53	55.6410	55.6384	28	49.1255	49.1275	4	34.8120	34.8107
52	55.5414	55.5407	27	48.6871	48.6894	3	34.0419	34.0416
51	55.4314	55.4296	26	48.2367	48.2376	2	33.2573	33.2588
50	55.3035	55.3050	25	47.7712	47.7721	1	32.4591	32.4621
49	55.1675	55.1669	24	47.2937	47.2928	0	31.6531	31.6517
48	55.0198	55.0153	23	46.8008	46.7997	P 1	29.9870	29.9894
47	54.8499	54.8502	22	46.2942	46.2929	2	29.1409	29.1376
46	54.6718	54.6715	21	45.7737	45.7722	3	28.2739	28.2720
45	54.4807	54.4793	20	45.2388	45.2378	4	27.3924	27.3927
44	54.2695	54.2736	19	44.6897	44.6897	5	26.4989	26.4996
43	54.0510	54.0543	18	44.1304	44.1277	6	25.5938	25.5927
42	53.8185	53.8214	17	43.5558	43.5519	7	24.6718	24.6720
41	53.5714	53.5748	16	42.9630	42.9623	8	23.7373	23.7376
40	53.3115	53.3147	15	42.3587	42.3590	9	22.7881	22.7895
39	53.0369	53.0410	14	41.7426	41.7418	10	21.8320	21.8276
38	52.7526	52.7536	13	41.1121	41.1108	11	20.8541	20.8520
37	52.4527	52.4526	12	40.4673	40.4661	12	19.8614	19.8627
36	52.1386	52.1379	11	39.8081	39.8074	13	18.8562	18.8596
35	51.8097	51.8095	10	39.1351	39.1350	14	17.8442	17.8429
34	51.4667	51.4675	9	38.4470	38.4488	15	16.8130	16.8125
33	51.1104	51.1118	8	37.7489	37.7488	16	15.7675	15.7684
32	50.7413	50.7423	7	37.0351	37.0350	17	14.7141	14.7106
31	50.3581	50.3592	6	36.3060	36.3073	18	13.6411	13.6391
30	49.9642	49.9624						

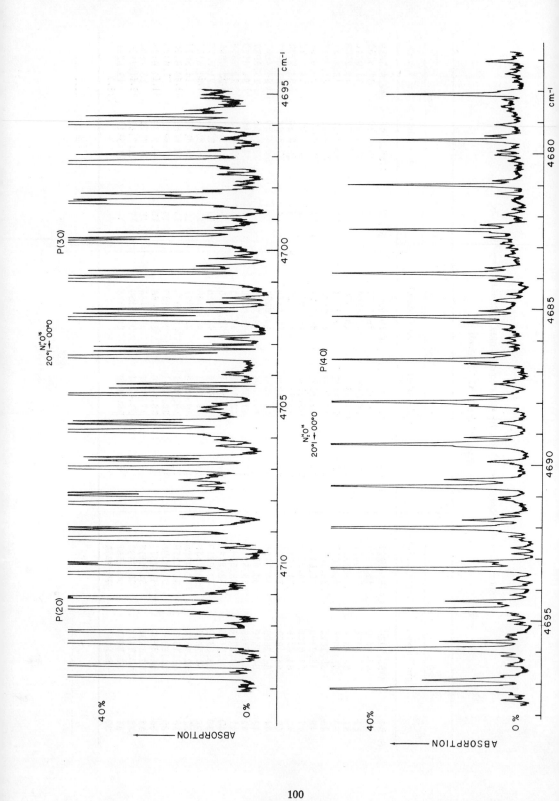

Vacuum Wave Numbers (cm⁻¹) for the Rotational Lines of the 20°1 —— 00°0 $(2\nu_1 + \nu_3)$
Band of the N$_2^{14}$O^{16} Molecule (Continued)

J	ν_{vac} cm⁻¹ (obs)	cm⁻¹ (calc)	J	ν_{vac} cm⁻¹ (obs)	cm⁻¹ (calc)
P 18	4713.6411	4713.6391	P 33	4695.9429	4695.9360
19	12.5532	12.5540	34	94.6486	94.6476
20	11.4583	11.4552	35	93.3439	93.3459
21	10.3436	10.3429	36	92.0403	92.0307
22	09.2164	09.2169	37	90.7034	90.7022
23	08.0751	08.0773	38	89.3601	89.3602
24	06.9268	06.9241	39	88.0070	88.0049
25	05.7582	05.7573	40	86.6349	86.6363
26	04.5756	04.5770	41	85.2516	85.2544
27	03.3862	03.3831	42	83.8572	83.8591
28	02.1765	02.1756	43	82.4523	82.4506
29	00.9531	00.9547	44	81.0270	81.0289
30	4699.7164	4699.7202	45	79.5924	79.5939
31	98.4746	98.4723	46	78.1472	78.1457
32	97.2100	97.2108			

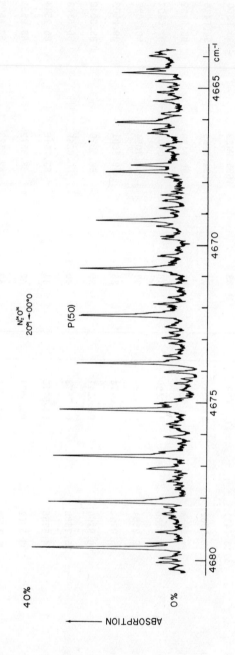

Vacuum Wave Numbers (cm⁻¹) for the Rotational Lines of the
20°1 ⟵ 00°0 ($2\nu_1 + \nu_3$) Band of the $N_2^{14}O^{16}$ Molecule (Continued)

J	ν_{vac}	
	cm⁻¹ (obs)	cm⁻¹ (calc)
P 45	4679.5924	4679.5939
46	78.1472	78.1457
47	76.6814	76.6843
48	75.2100	75.2097
49	73.7235	73.7220
50	72.2198	72.2212
51	70.7057	70.7072
52	69.1832	69.1802
53	67.6384	67.6401
54	66.0835	66.0870
55	64.5235	64.5209

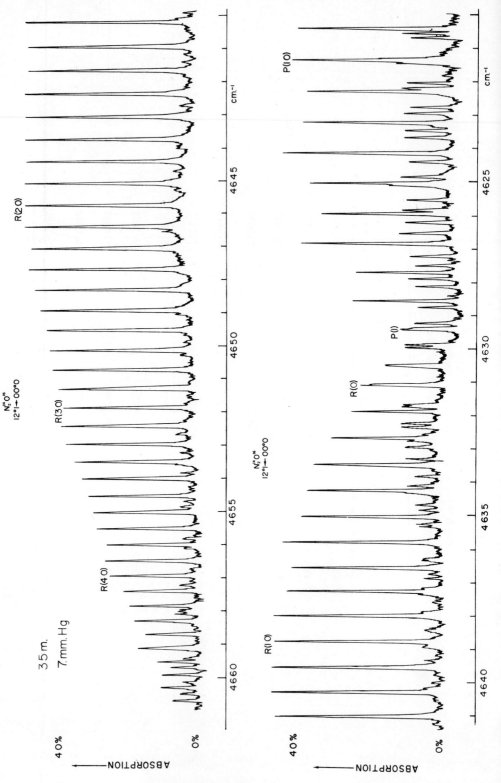

104

Vacuum Wave Numbers (cm⁻¹) for the Rotational Lines of the 12°1 ⟵ 00°0 ($\nu_1 + 2\nu_2 + \nu_3$)

Band of the N₂¹⁴O¹⁶ Molecule

J	ν_{vac} cm⁻¹ (obs)	cm⁻¹ (calc)	J	ν_{vac} cm⁻¹ (obs)	cm⁻¹ (calc)	J	ν_{vac} cm⁻¹ (obs)	cm⁻¹ (calc)
R 48	4660.3861	4660.3805	R 28	4650.7070	4650.7045	R 8		4637.3239
47	59.9942	59.9908	27	50.1215	50.1205	7		36.5628
46	59.5910	59.5908	26	49.5269	49.5275	6	4635.7924	35.7931
45	59.1767	59.1806	25		48.9252	5	35.0148	35.0149
44	58.7574	58.7603	24		48.3139	4	34.2246	34.2281
43		58.3299	23		47.6936	3		33.4328
42		57.8895	22		47.0642	2		32.6288
41		57.4391	21		46.4258	1		31.8164
40		56.9789	20	45.7790	45.7785	0		30.9954
39		56.5088	19	45.1245	45.1223	P 1	4629.3216	4629.3279
38	56.0308	56.0289	18	44.4568	44.4572	2		28.4814
37	55.5402	55.5393	17		43.7833	3		27.6263
36	55.0415	55.0402	16		43.1005	4		26.7628
35	54.5307	54.5312	15		42.4089	5	25.8920	25.8907
34	54.0103	54.0128	14		41.7086	6	25.0082	25.0101
33		53.4848	13	40.9999	40.9995	7	24.1190	24.1210
32		52.9475	12	40.2833	40.2817	8		23.3234
31		52.4007	11	39.5537	39.5552	9		22.3173
30		51.8446	10		38.8201	10		21.4027
29		51.2792	9		38.0763	11	20.4782	20.4795

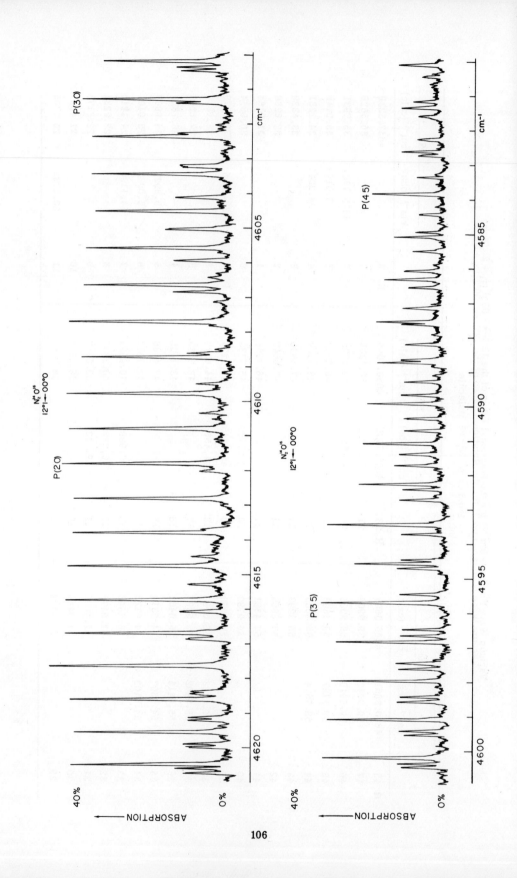

Vacuum Wave Numbers (cm⁻¹) for the Rotational Lines of the 12°1 ⟵ 00°0 ($\nu_1 + 2\nu_2 + \nu_3$)

Band of the N₂¹⁴O¹⁶ Molecule (Continued)

J	ν_{vac} cm⁻¹ (obs)	cm⁻¹ (calc)	J	ν_{vac} cm⁻¹ (obs)	cm⁻¹ (calc)
P 11	4620.4782	4620.4795	P 30		4601.2972
12	19.5460	19.5477	31		00.1996
13		18.6075	32	4599.0940	4599.0931
14		17.6586	33	97.9752	97.9774
15		16.7012	34		96.8528
16	15.7375	15.7352	35		95.7189
17	14.7617	14.7605	36		94.5760
18	13.7736	13.7773	37	93.4307	93.4238
19		12.7854	38	92.2602	92.2624
20		11.7848	39		91.0916
21		10.7755	40		89.9116
22	09.7592	09.7576	41		88.7223
23	08.7298	08.7309	42	87.5235	87.5233
24		07.6954	43	86.3100	86.3149
25		06.6512	44		85.0970
26		05.5981	45		83.8695
27	04.5376	04.5362	46		82.6324
28	03.4643	03.4655	47	81.3811	81.3856
29		02.3858			

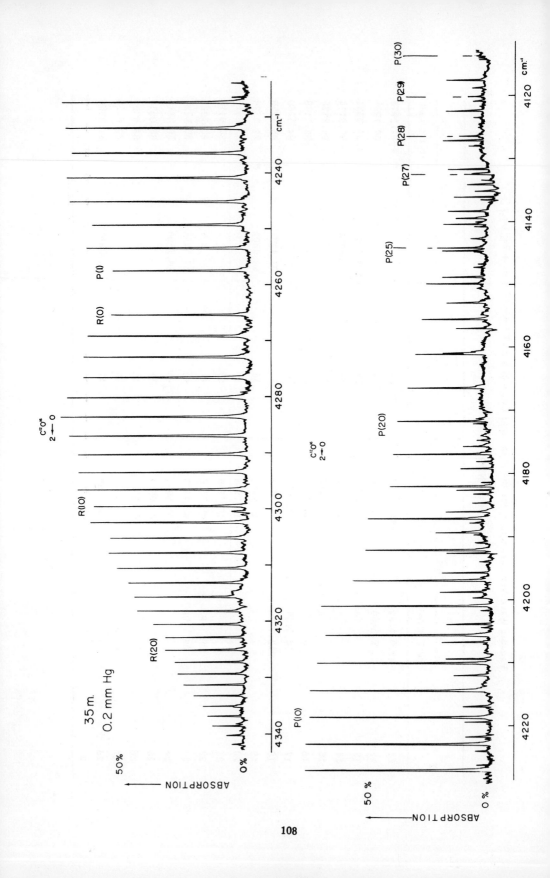

Vacuum Wave Numbers (cm^{-1}) for the Rotational Lines of the 2 ←— 0 Band of the C^{12}O^{16} Molecule

J	ν_{vac} cm^{-1} (calc)	J	ν_{vac} cm^{-1} (calc)	J	ν_{vac} cm^{-1} (calc)	J	ν_{vac} cm^{-1} (calc)
R 30	4343.8061	R 14	4309.2558	P 1	4256.2196	P 16	4190.2424
29	42.1996	13	06.4764	2	52.3047	17	85.2971
28	40.5187	12	03.6250	3	48.3201	18	80.2843
27	38.7635	11	00.7018	4	44.2659	19	75.2041
26	36.9342	10	4297.7065	5	40.1423	20	70.0568
25	35.0310	9	94.6397	6	35.9494	21	64.8423
24	33.0540	8	91.5014	7	31.6874	22	59.5609
23	31.0033	7	88.2918	8	27.3564	23	54.2128
22	28.8790	6	85.0111	9	22.9565	24	48.7980
21	26.6814	5	81.6592	10	18.4880	25	43.3167
20	24.4106	4	78.2365	11	13.9509	26	37.7691
19	22.0666	3	74.7430	12	09.3454	27	32.1553
18	19.6498	2	71.1790	13	04.6716	28	26.4755
17	17.1601	1	67.5445	14	4199.9298	29	20.7297
16	14.5978	0	63.8396	15	95.1200	30	14.9183
15	11.9630						

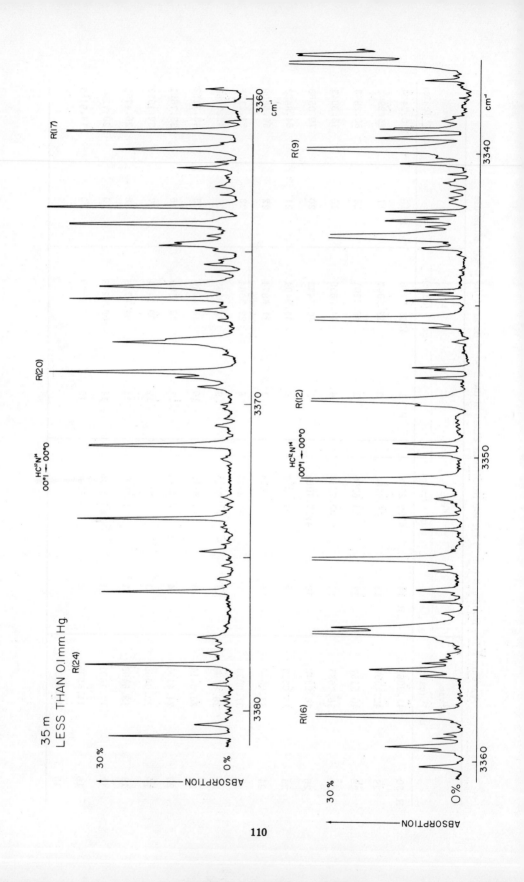

35 m
LESS THAN 0.1 mm Hg

$HC^{12}N^{14}$
$00^{0}1 \leftarrow 00^{0}0$

$HC^{12}N^{14}$
$00^{0}1 \leftarrow 00^{0}0$

R(17)
R(20)
R(24)

R(9)
R(12)
R(16)

ABSORPTION
30%
0%

ABSORPTION
30%
0%

cm⁻¹
3360
3370
3380

cm⁻¹
3340
3350
3360

Vacuum Wave Numbers (cm⁻¹) for the Rotational Lines of the

$00°1 \leftarrow 00°0$ (ν_3) Band of the $HC^{12}N^{14}$ Molecule

J	ν_{vac}	
	cm⁻¹ (obs)	cm⁻¹ (calc)
R 25	3380.8356	3380.8366
24	78.4450	78.4430
23		76.0271
22		73.5893
21	71.1277	71.1283
20		68.6455
19		66.1405
18	63.6128	63.6134
17	61.0660	61.0643
16	58.4947	58.4931
15	55.8978	55.9001
14	53.2847	53.2851
13	50.6465	50.6483
12	47.9866	47.9898
11		45.3096
10		42.6077
9	39.8845	39.8843
8	37.1369	37.1393

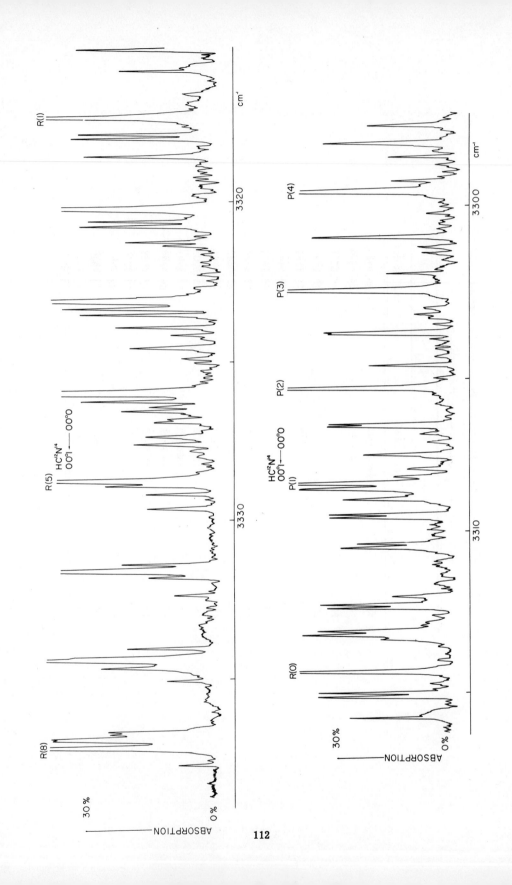

Vacuum Wave Numbers (cm⁻¹) for the Rotational Lines of the
00°1 ⟵ 00°0 (ν_3) Band of the $HC^{12}N^{14}$ Molecule (Continued)

J		ν_{vac}	
		cm⁻¹ (obs)	cm⁻¹ (calc)
R	8	3337.1369	3337.1393
	7		34.3729
	6		31.5852
	5		28.7761
	4		25.9457
	3	23.0945	23.0942
	2		20.2216
	1	17.3260	17.3279
	0	14.4144	14.4132
P	1	08.5215	08.5212
	2	05.5438	05.5440
	3	02.5457	02.5461
	4		3299.5275

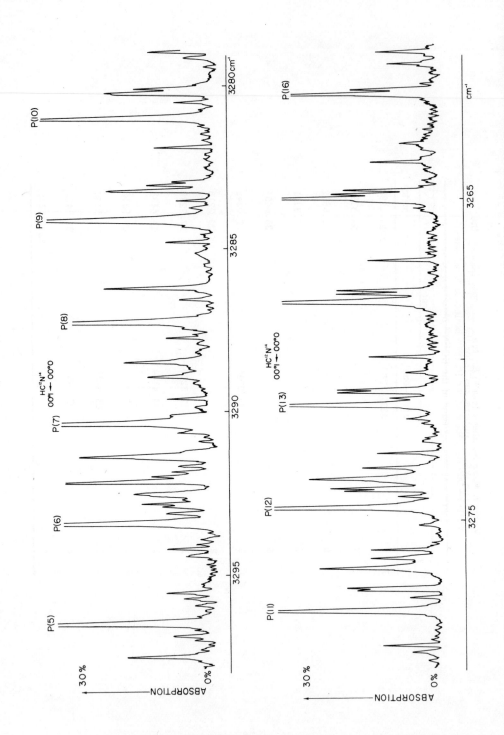

Vacuum Wave Numbers (cm⁻¹) for the Rotational Lines of the

$00^\circ 1 \longleftarrow 00^\circ 0$ (ν_3) Band of the $HC^{12}N^{14}$ Molecule (Continued)

J	ν_{vac}	
	cm⁻¹ (obs)	cm⁻¹ (calc)
P 5	3296.4898	3296.4885
6	93.4265	93.4289
7	90.3474	90.3489
8	87.2475	87.2485
9	84.1292	84.1279
10	80.9884	80.9871
11	77.8228	77.8262
12	74.6452	74.6453
13	71.4475	71.4443
14	68.2312	68.2235
15	64.9795	64.9829
16		61.7225

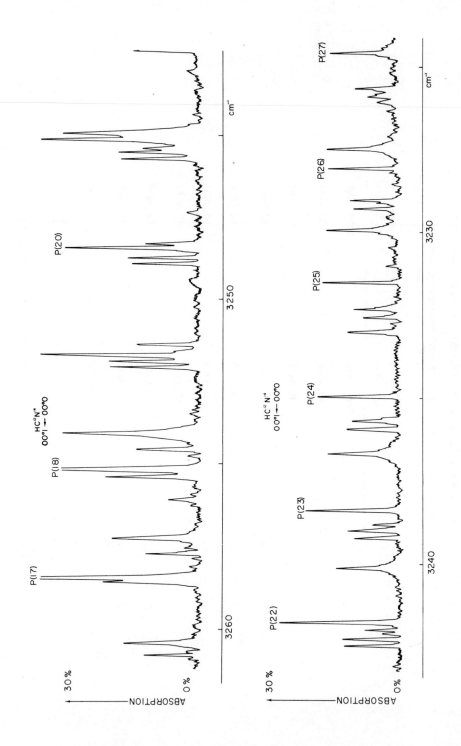

Vacuum Wave Numbers (cm⁻¹) for the Rotational Lines of the
00°1 ⟵ 00°0 (ν_3) Band of the HC^{12}N^{14} Molecule (Continued)

J	ν_{vac}	
	cm⁻¹ (obs)	cm⁻¹ (calc)
P 17		3258.4421
18	3255.1439	55.1429
19	51.8220	51.8238
20	48.4828	48.4853
21	45.1257	45.1274
22	41.7491	41.7503
23	38.3505	38.3535
24	34.9359	34.9386
25	31.5033	31.5041
26	28.0487	28.0508
27	24.5781	24.5786

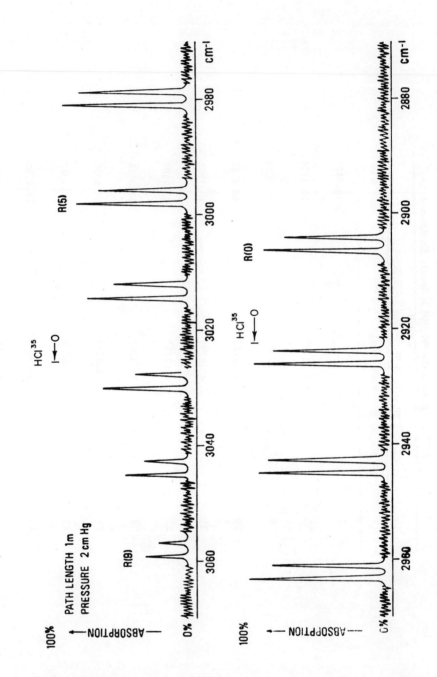

Vacuum Wave Numbers (cm^{-1}) for the Rotational Lines of the
1 ⟵ 0 Band of the HCl^{35} Molecule[a]

J	ν_{vac}	
	cm^{-1} (obs)	cm^{-1} (calc)
R 15	3129.3031	3129.3042
14	19.5362	19.5413
13	09.0026	09.0050
12	3097.7034	3097.7048
11	85.6539	85.6502
10	72.8490	72.8509
9	59.3179	59.3171
8	45.0569	45.0592
7	30.0862	30.0876
6	14.4114	14.4134
5	2998.0438	2998.0476
4	81.0013	81.0015
3	63.2865	63.2866
2	44.9154	44.9146
1	25.8950	25.8977
0	06.2521	06.2479

[a] The HCl lines between R 15 and R 9 can be observed in absorption by using longer pathlengths or higher pressures. Although the figure does not include these lines, the spectral positions are given in the table because these lines have actual- by been observed in absorption, and measured.

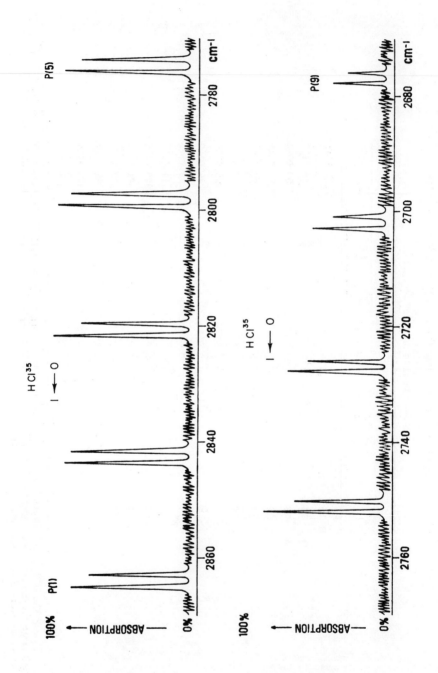

Vacuum Wave Numbers (cm⁻¹) for the Rotational Lines of the
1⟶0 Band of the HCl³⁵ Molecule [a] (Continued)

J	ν_{vac}	
	cm⁻¹ (obs)	cm⁻¹ (calc)
P 1	2865.0967	2865.0991
2	43.6234	43.6254
3	21.5713	21.5691
4	2798.9401	2798.9433
5	75.7602	75.7609
6	52.0363	52.0353
7	27.7774	27.7797
8	03.0068	03.0074
9	2677.7320	2677.7320
10	51.9664	51.9668
11	25.7272	25.7255
12	2599.0208	2599.0216
13	71.8703	71.8686
14	44.2817	44.2801
15	16.2724	16.2696

[a] The HCl lines between P 9 and P 15 can be observed in absorption by using longer pathlengths or higher pressures. Although the figure does not include these lines, the spectral positions are given in the table because these lines have actually been observed in absorption, and measured.

121

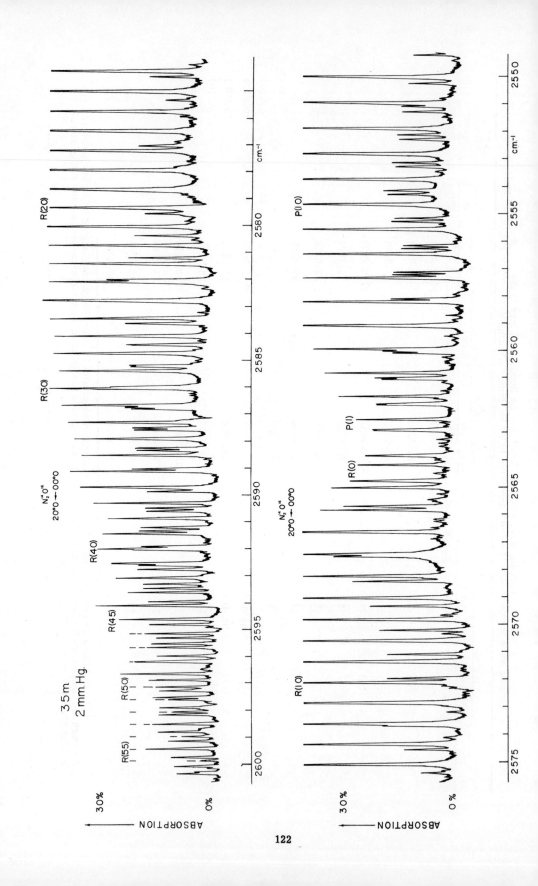

Vacuum Wave Numbers (cm⁻¹) for the Rotational Lines of the 20°0←00°0 ($2\nu_1$) Band of the $N_2^{14}O^{16}$ Molecule

J	ν_{vac} cm⁻¹ (obs)	cm⁻¹ (calc)	J	ν_{vac} cm⁻¹ (obs)	cm⁻¹ (calc)	J	ν_{vac} cm⁻¹ (obs)	cm⁻¹ (calc)
R 56	2599.8761	2599.8749	R 32	2587.1650	2587.1657	R 8	2570.5753	2570.5759
55		99.4211	31	86.5523	86.5526	7		69.7993
54	98.9607	98.9608	30	85.9308	85.9329	6	69.0165	69.0159
53		98.4939	29	85.3055	85.3063	5		68.2257
52	98.0192	98.0206	28	84.6731	84.6730	4	67.4306	67.4286
51	97.5380	97.5407	27	84.0335	84.0328	3		66.6247
50	97.0527	97.0543	26	83.3860	83.3859	2		65.8140
49		96.5614	25	82.7354	82.7322	1	64.9970	64.9964
48		96.0619	24	82.0725	82.0717	0		64.1720
47	95.5552	95.5557	23	81.4046	81.4044	P 1	62.5020	62.5028
46	95.0410	95.0430	22	80.7318	80.7303	2		61.6579
45	94.5226	94.5237	21	80.0491	80.0494	3	60.8078	60.8063
44		93.9978	20		79.3616	4	59.9476	59.9479
43	93.4637	93.4652	19	78.6672	78.6670	5		59.0826
42	92.9235	92.9259	18	77.9649	77.9657	6	58.2084	58.2106
41	92.3806	92.3801	17	77.2581	77.2574	7	57.3318	57.3318
40	91.8263	91.8275	16	76.5420	76.5424	8		56.4462
39	91.2669	91.2682	15		75.8205	9		55.5539
38	90.7029	90.7023	14	75.0924	75.0918	10		54.6547
37	90.1296	90.1297	13	74.3555	74.3562	11	53.7494	53.7488
36	89.5521	89.5503	12	73.6148	73.6139	12	52.8375	52.8362
35		88.9643	11	72.8626	72.8645	13	51.9179	51.9168
34	88.3708	88.3715	10		72.1086	14	50.9910	50.9907
33	87.7716	87.7719	9	71.3463	71.3457	15		50.0579

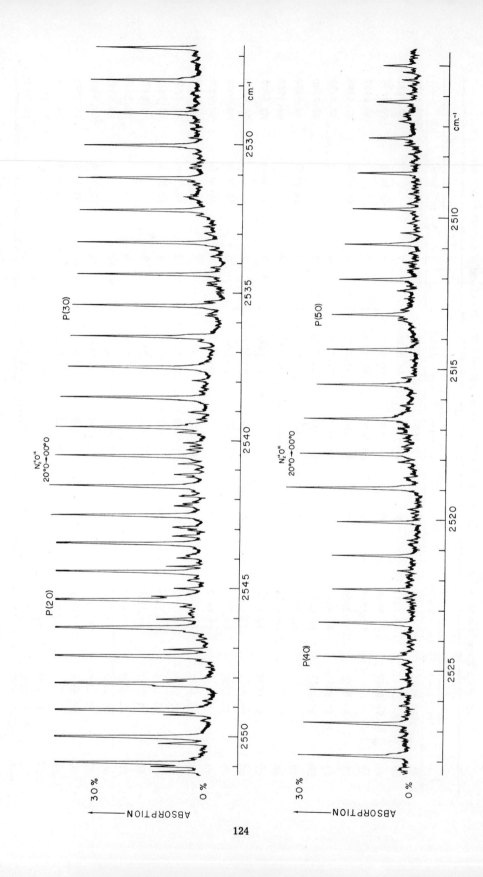

Vacuum Wave Numbers (cm⁻¹) for the Rotational Lines of the 20°0 ←—00°0 (2ν_1)

Band of the N$_2^{14}$O^{16} Molecule (Continued)

J	ν_{vac} cm⁻¹ (obs)	ν_{vac} cm⁻¹ (calc)	J	ν_{vac} cm⁻¹ (obs)	ν_{vac} cm⁻¹ (calc)	J	ν_{vac} cm⁻¹ (obs)	ν_{vac} cm⁻¹ (calc)
P 14	2550.9910	2550.9907	P 29	2536.2983	2536.2981	P 44	2520.1336	2520.1335
15		50.0579	30		35.2657	45		19.0048
16	49.1197	49.1184	31	34.2282	34.2268	46		17.8698
17	48.1734	48.1721	32		33.1813	47	16.7281	16.7286
18		47.2192	33		32.1294	48		15.5812
19	46.2594	46.2595	34	31.0714	31.0709	49	14.4281	14.4276
20		45.2932	35		30.0060	50		13.2677
21	44.3218	44.3203	36	28.9358	28.9347	51		12.1017
22	43.3419	43.3406	37	27.8564	27.8569	52	10.9285	10.9296
23		42.3544	38		26.7727	53		09.7513
24	41.3611	41.3615	39	25.6826	25.6820	54		08.5669
25		40.3620	40		24.5850	55		07.3765
26	39.3579	39.3559	41		23.4816	56		06.1800
27	38.3429	38.3432	42	22.3707	22.3719	57	04.9790	04.9774
28		37.3239	43		21.2558			

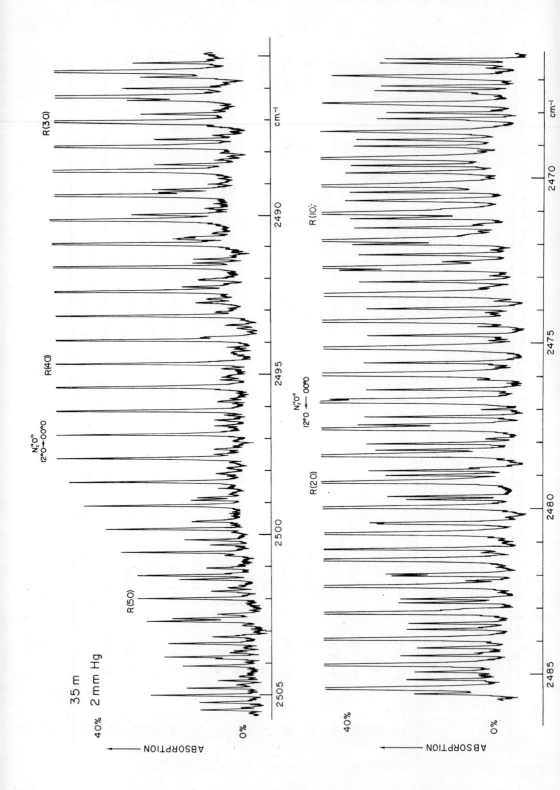

Vacuum Wave Numbers (cm⁻¹) for the Rotational Lines of the 12°0 ←——00°0 ($\nu_1 + 2\nu_2$) Band of the $N_2^{14}O^{16}$ Molecule

J	ν_{vac} cm⁻¹ (obs)	ν_{vac} cm⁻¹ (calc)	J	ν_{vac} cm⁻¹ (obs)	ν_{vac} cm⁻¹ (calc)	J	ν_{vac} cm⁻¹ (obs)	ν_{vac} cm⁻¹ (calc)
R 55	2505.4504	2505.4492	R 38	2493.1399	2493.1394	R 21	2479.9725	2479.9734
54	04.7547	04.7526	37	92.3859	92.3863	20	79.1762	79.1771
53	04.0543	04.0523	36	91.6274	91.6304	19	78.3786	78.3786
52	03.3475	03.3485	35		90.8716	18	77.5790	77.5780
51	02.6381	02.6412	34		90.1099	17	76.7736	76.7753
50	01.9311	01.9304	33	89.3449	89.3455	16	75.9695	75.9705
49	01.2169	01.2160	32	88.5764	88.5784	15	75.1591	75.1637
48	2500.4970	2500.4983	31	87.8113	87.8087	14		74.3549
47		2499.7771	30	87.0350	87.0363	13	73.5431	73.5441
46	2499.0539	99.0525	29	86.2612	86.2613	12	72.7311	72.7313
45	98.3249	98.3246	28	85.4810	85.4838	11	71.9175	71.9167
44		97.5934	27		84.7037	10		71.1001
43		96.8589	26	83.9184	83.9212	9	70.2808	70.2817
42	96.1213	96.1213	25	83.1374	83.1363	8	69.4625	69.4614
41	95.3781	95.3804	24	82.3469	82.3490	7	68.6376	68.6393
40	94.6356	94.6365	23		81.5594	6	67.8139	67.8154
39	93.8913	93.8894	22	80.7648	80.7675	5	66.9896	66.9896

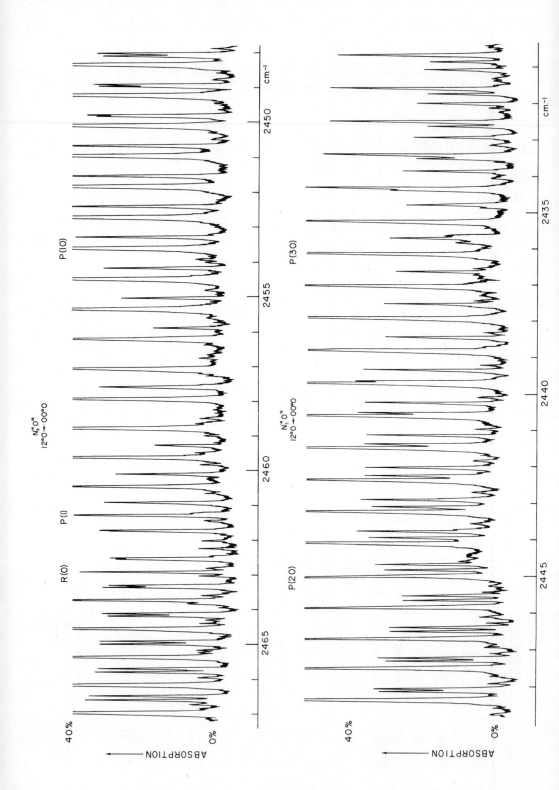

128

Vacuum Wave Numbers (cm⁻¹) for the Rotational Lines of the 12°0 ⟵ 00°0 ($\nu_1 + 2\nu_2$)

Band of the N$_2^{14}$O^{16} Molecule (Continued)

J	ν_{vac} cm⁻¹ (obs)	cm⁻¹ (calc)	J	ν_{vac} cm⁻¹ (obs)	cm⁻¹ (calc)	J	ν_{vac} cm⁻¹ (obs)	cm⁻¹ (calc)
R 5	2466.9896	2466.9896	P 9	2454.3965	2454.3938	P 23	2442.2777	2442.2784
4	66.1604	66.1621	10	53.5419	53.5402	24	41.4024	41.3987
3	65.3344	65.3328	11	52.6843	52.6849	25	40.5182	40.5170
2	64.5029	64.5018	12	51.8272	51.8278	26	39.6346	39.6332
1	63.6697	63.6689	13	50.9682	50.9689	27	38.7457	38.7473
0	62.8345	62.8344	14	50.1084	50.1082	28	37.8584	37.8592
P 1	61.1590	61.1601	15	49.2453	49.2457	29	36.9684	36.9690
2	60.3199	60.3203	16	48.3839	48.3814	30	36.0816	36.0765
3	59.4780	59.4789	17	47.5177	47.5153	31	35.1815	35.1818
4	58.6363	58.6357	18	46.6458	46.6473	32	34.2850	34.2848
5	57.7909	57.7908	19	45.7774	45.7774	33	33.3877	33.3855
6	56.9455	56.9441	20		44.9055	34	32.4820	32.4837
7	56.0969	56.0958	21	44.0328	44.0318	35	31.5792	31.5796
8	55.2450	55.2456	22	43.1585	43.1561	36	30.6747	30.6731

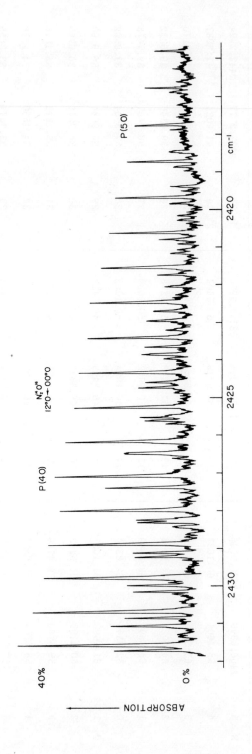

Vacuum Wave Numbers (cm^{-1}) for the Rotational Lines of the 12°0 $\longleftarrow$ 00°0 ($\nu_1 + 2\nu_2$)
Band of the N$_2^{14}$O^{16} Molecule (Continued)

J	ν_{vac}		J	ν_{vac}	
	cm^{-1} (obs)	cm^{-1} (calc)		cm^{-1} (obs)	cm^{-1} (calc)
P 35	2431.5792	2431.5796	P 47	2419.5897	2420.5295
36	30.6747	30.6731	48		19.5907
37	29.7613	29.7642	49		18.6489
38	28.8514	28.8527	50		17.7041
39	27.9389	27.9387	51		16.7564
40		27.0221	52		15.8056
41		26.1028	53		14.8517
42	25.1817	25.1809	54		13.8948
43	24.2589	24.2562	55		12.9349
44		23.3288	56		11.9718
45	22.3978	22.3986	57		11.0055
46	21.4666	21.4655			

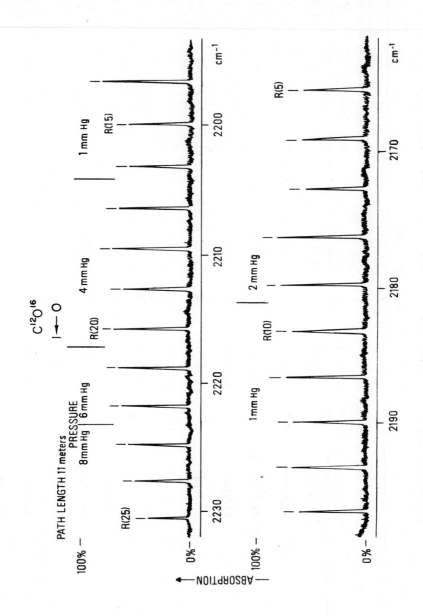

$C^{12}O^{16}$

1 ← 0

PATH LENGTH 11 meters

PRESSURE

ABSORPTION

Vacuum Wave Numbers (cm^{-1}) for the Rotational Lines of the 1 ← 0 Band of the C^{12}O^{16} Molecule

J	ν_{vac} cm^{-1} (calc)	J	ν_{vac} cm^{-1} (calc)
R 30	2244.3757	R 17	2206.3540
29	41.6841	16	03.1616
28	38.9531	15	2199.9317
27	36.1829	14	96.6645
26	33.3735	13	93.3601
25	30.5252	12	90.0187
24	27.6331	11	86.6403
23	24.7124	10	83.2251
22	21.7481	9	79.7733
21	18.7454	8	76.2850
20	15.7045	7	72.7604
19	12.6256	6	69.1996
18	09.5087	5	65.6028

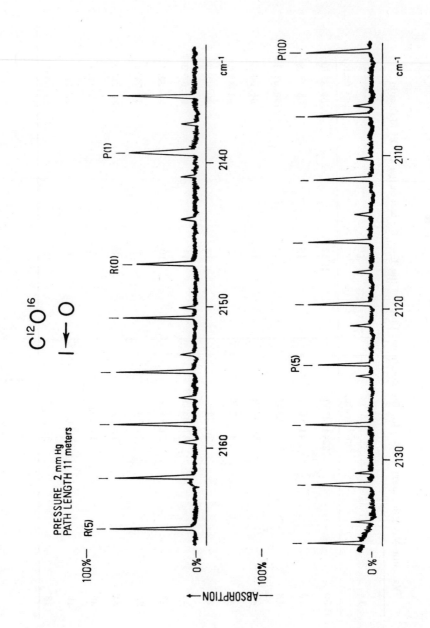

$C^{12}O^{16}$

$I \longleftarrow O$

PRESSURE 2 mm Hg
PATH LENGTH 11 meters

R(5)   R(0)   P(1)

P(5)   P(10)

ABSORPTION

2140   2150   2160   cm⁻¹

2110   2120   2130   cm⁻¹

100%   0%   100%   0%

134

Vacuum Wave Numbers (cm^{-1}) for the Rotational Lines of the 1 ⟵ 0
Band of the C^{12}O^{16} Molecule (Continued)

J	ν_{vac} cm^{-1} (calc)	J	ν_{vac} cm^{-1} (calc)
R 5	2165.6028	P 3	2131.6336
4	61.9700	4	27.6844
3	58.3016	5	23.7008
2	54.5975	6	19.6829
1	50.8579	7	15.6309
0	47.0831	8	11.5449
P 1	39.4281	9	07.4251
2	35.5482	10	03.2715

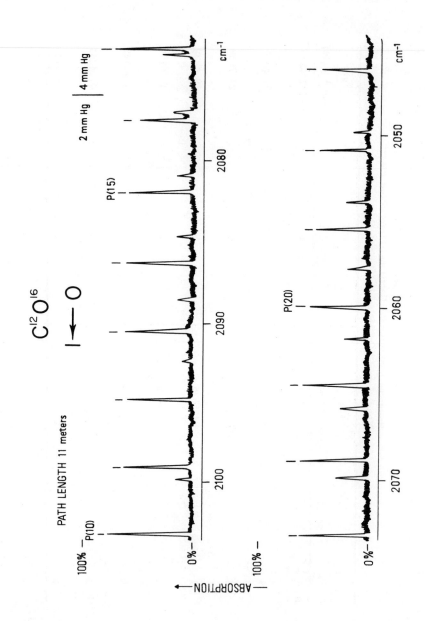

Vacuum Wave Numbers (cm^{-1}) for the Rotational Lines of the $1 \leftarrow 0$ Band of the C^{12}O^{16} Molecule (Continued)

J	ν_{vac} cm^{-1} (calc)	J	ν_{vac} cm^{-1} (calc)
P 10	2103.2715	P 21	2055.4011
11	2099.0845	22	50.8547
12	94.8640	23	46.2766
13	90.6103	24	41.6668
14	86.3234	25	37.0255
15	82.0037	26	32.3529
16	77.6511	27	27.6491
17	73.2658	28	22.9142
18	68.8480	29	18.1484
19	64.3979	30	13.3519
20	59.9155		

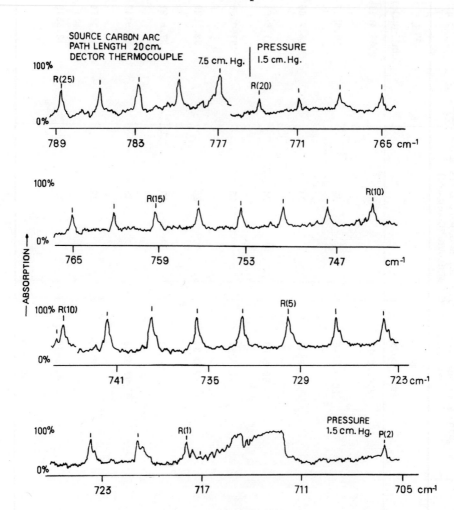

SOURCE CARBON ARC
PATH LENGTH 20 cm.
DECTOR THERMOCOUPLE

7.5 cm. Hg.

PRESSURE
1.5 cm. Hg.

100%

R(25)

0%

789   783   777

R(20)

771   765 cm⁻¹

100%

R(15)

R(10)

0%

765   759   753   747   cm⁻¹

ABSORPTION →

100% R(10)

R(5)

0%

741   735   729   723 cm⁻¹

PRESSURE
1.5 cm. Hg.

100%

R(1)

P(2)

0%

723   717   711   705 cm⁻¹

Vacuum Wave Numbers (cm^{-1}) for the Rotational Lines of the $01^10 \leftarrow 00^00$ (ν_2) Band of the HC^{12}N^{14} Molecule

J	ν_{vac}* (cm^{-1})	J	ν_{vac}* (cm^{-1})
R 25	788.491	R 12	750.362
24	785.572	11	747.415
23	782.651	10	744.467
22	779.727	9	741.518
21	776.800	8	738.568
20	773.871	7	735.617
19	770.940	6	732.665
18	768.006	5	729.712
17	765.070	4	726..758
16	762.132	3	723.804
15	759.192	2	720.849
14	756.250	1	717.894
13	753.307	P 2	706.070

* Values quoted are "calculated" values. Accuracy is believed to be no better than ± 0.02 cm^{-1}.

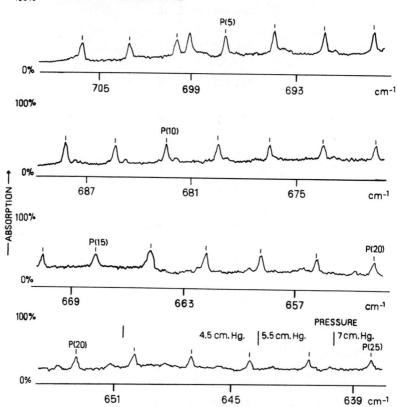

SOURCE NERNST GLOWER
PATH LENGTH 20 cm.
DETECTOR THERMOCOUPLE

Vacuum Wave Numbers (cm⁻¹) for the Rotational Lines of the $01^10 \leftarrow 00^00$ (ν_2) Band of the $HC^{12}N^{14}$ Molecule

J	ν_{vac}^* (cm⁻¹)	J	ν_{vac}^* (cm⁻¹)
P 2	706.070	P 14	670.595
3	703.112	15	667.641
4	700.155	16	664.687
5	697.199	17	661.734
6	694.242	18	658.782
7	691.285	19	655.830
8	688.329	20	652.878
9	685.372	21	649.927
10	682.416	22	646.977
11	679.460	23	644.027
12	676.505	24	641.078
13	673.550	25	638.129

*Values quoted are "calculated" values. Accuracy is believed to be no better than ± 0.02 cm⁻¹.

CHAPTER IV
Absorption Standards in the Far Infrared

Pure Rotational Lines of Water Vapor in the Region 16-200 μ (600-50 cm⁻¹)

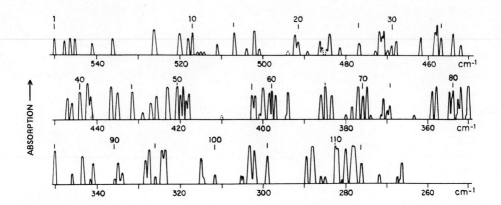

The spectra sketched above were obtained in the first order of a 1000 lines per inch Bausch and Lomb plane replica grating installed in a 1 m focal length Pfund-type vacuum spectrograph with a CsBr fore-prism. The pressure of water vapor used was 8 mm of Hg in a path of 2 m. A Nernst glower provided a source of continuous radiation and a Charles M. Reeder thermocouple equipped with a CsBr window was employed to detect the infrared radiation.

Measured Positions of the Pure Rotational Lines of the H_2O^{16} Molecule

Serial[a] no.	$\nu(\text{cm}^{-1})_{\text{vac}}$	Serial[a] no.	$\nu(\text{cm}^{-1})_{\text{vac}}$	Serial[a] no.	$\nu(\text{cm}^{-1})_{\text{vac}}$
1	550.00	40	443.70	79	354.60
2	47.81	41	42.09	80	54.19
3	46.30	42	41.75	81	51.98
4	45.29	43	36.46	82	51.80
5	41.07	44	34.82	83	51.21
6	36.25	45	31.16	84	50.50
7	25.97	46	28.85	85	49.77
8	19.60	47	26.33	86	45.84
9	17.77	48	25.34	87	43.21
10	16.81	49	23.03	88	41.27
11	15.89	50	19.88	89	40.55
12	15.05	51	19.13	90	35.68
13	14.28	52	18.52	91	35.16
14	10.54	53	17.90	92	34.63
15	06.94	54	17.70	93	28.16
16	04.41	55	00.49	94	27.60
17	02.26	56	00.25	95	25.78
18	01.57	57	399.51	96	23.92
19	492.05	58	98.97	97	23.67
20	91.63	59	97.69	98	15.08
21	89.56	60	97.34	99	14.74
22	86.12	61	96.44	100	11.72
23	83.97	62	94.63	101	04.87
24	81.05	63	94.28	102	04.55
25	76.38	64	85.54	103	03.02
26	72.75	65	84.88	104	01.87
27	72.39	66	83.82	105	298.42
28	72.20	67	80.98	106	90.73
29	70.52	68	78.55	107	89.46
30	68.75	69	76.27	108	85.57
31	67.93	70	75.38	109	84.78
32	61.45	71	74.54	110	82.72
33	57.96	72	73.88	111	82.00
34	57.73	73	70.02	112	80.34
35	56.87	74	69.61	113	78.47
36	52.89	75	69.36	114	77.94
37	51.72	76	62.75	115	76.15
38	46.93	77	58.49	116	71.86
39	46.36	78	57.27		

[a] Serial numbers refer to the chart on the opposite page.
Accuracy of measurements on single lines is believed to be no better than ±0.03 cm^{-1}.

H₂O¹⁶

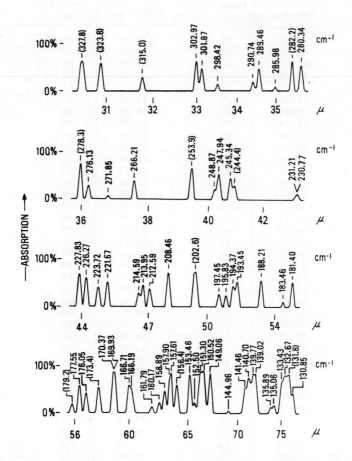

Pure rotational lines of H_2O^{16} recorded with a Perkin-Elmer Model 301 spectrophotometer. The wave number value given for each line refers to vacuum. The spectrum shown above has been recorded by keeping the grating in air.

The first 14 lines between 30 and 37 μ have already been included in the table on p. 145. The measurements reported in the table should be considered more accurate than the values appearing in the above figure which have been derived purely from energy levels. A comparison of the overlap region between the data presented in the table and those shown above may provide an idea of the precision of data appearing in the figures on pp. 146-147.

H₂O¹⁶

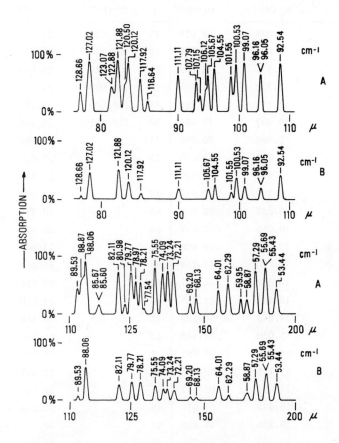

Pure rotational lines of H_2O^{16} recorded with a Perkin-Elmer Model 301 spectrophotometer. The wave number value given for each line refers to vacuum. A. Spectrophotometer flushed with dry nitrogen. B. Spectrophotometer flushed with dry nitrogen and P_2O_5 trays kept inside the instrument for 12 hours before observations were recorded.

Refer to the remarks in the second paragraph of the caption for the figure on p. 146.

Vacuum Wave Numbers (cm^{-1}) for the Pure Rotational Lines of the
$C^{12}O^{16}$, $N_2^{14}O^{16}$, and $HC^{12}N^{14}$ Molecules

J	$C^{12}O^{16}$ [a]	$N_2^{14}O^{16}$ [a]	$HC^{12}N^{14}$ [b]	J	$C^{12}O^{16}$ [a]	$N_2^{14}O^{16}$ [a]	$HC^{12}N^{14}$ [b]
0	3.845	0.838	2.956	25	99.541	21.776	76.663
1	7.690	1.676	5.913	26	103.334	22.613	79.595
2	11.534	2.514	8.869	27	107.124	23.449	82.524
3	15.379	3.352	11.825	28	110.909	24.285	85.453
4	19.222	4.190	14.781	29	114.690	25.122	88.379
5	23.065	5.028	17.736	30	118.467	25.958	91.30
6	26.907	5.866	20.691	31		26.794	94.22
7	30.748	6.704	23.646	32		27.629	97.14
8	34.588	7.542	26.599	33		28.465	100.06
9	38.426	8.380	29.553	34		29.301	102.98
10	42.263	9.217	32.505	35		30.136	105.89
11	46.098	10.055	35.457	36		30.971	108.80
12	49.932	10.893	38.408	37		31.806	111.71
13	53.763	11.730	41.358	38		32.641	114.61
14	57.593	12.568	44.307	39		33.476	117.51
15	61.420	13.405	47.255	40		34.310	120.41
16	65.245	14.243	50.202	41		35.145	
17	69.068	15.080	53.148	42		35.979	
18	72.888	15.918	56.092	43		36.813	
19	76.705	16.755	59.036	44		37.647	
20	80.519	17.592	61.977	45		38.481	
21	84.330	18.429	64.918	46		39.314	
22	88.138	19.266	67.856	47		40.147	
23	91.943	20.103	70.793	48		40.980	
24	95.744	20.940	73.729	49		41.813	
				50		42.646	

[a] To obtain the pure rotational absorption spectra of $C^{12}O^{16}$ and $N_2^{14}O^{16}$, Palik and Rao [E. D. Palik and K. Narahari Rao, *J. Chem. Phys.* 25, 1174 (1956)] used a pressure of 40-60 cm of Hg in an absorption cell 43 cm in length.

[b] For recording the pure rotational spectrum of HCN, an absorption path of 30-60 cm at a pressure of 20-30 mm of Hg has been found to be adequate. [H. A. Gebbie, National Physical Laboratory, Teddington, England, private communication (1965).]

CHAPTER V

TECHNIQUES EMPLOYED FOR WAVELENGTH CALIBRATIONS IN THE INFRARED

A. INTRODUCTORY COMMENTS

This chapter presents a résumé of some of the techniques currently used for determining spectral positions in the infrared. The primary purpose of providing this résumé is to illustrate the usefulness of the standards listed in the previous chapters. Undoubtedly, other techniques will be and should be explored, and the problem of precision in determining wavelengths and wave numbers in the infrared is expected to continue to engage the attention of infrared spectroscopists for many years in the future.

Owing to freedom from the physical limitations of refracting components and to adaptability for use in scanning systems, plane gratings are used extensively as dispersing elements for the observation of high-resolution infrared spectra in the region of the spectrum which extends from 1 to about 1000 μ. In obtaining good wavelength calibrations in the infrared, one is faced with problems different from those encountered in other regions of the electromagnetic spectrum: For example, in the microwave region where frequencies are measured directly, high precision can be obtained. Also, in the regions in which photographic techniques can be employed, it is possible to record on a photographic plate, both the unknown spectrum as well as the standards, and the wavelengths of the spectral lines can then be determined by interpolation between the standards. In the infrared region, the scanning of the spectra is mostly accomplished by rotating the grating. Therefore, the accuracy of the data recorded depends, to some extent, upon the smoothness and precision of the mechanical drive which rotates the grating, and also on the techniques employed. In fact, prior to the year 1950, the usefulness of some of the high-resolution infrared data suffered from limitations resulting from inadequacies in the precision with which spectral positions could be determined.

In attempting to attain high precision in the wavelength determinations of spectral lines, it is important to be cognizant of the various possible sources of error. A rather serious limitation to precision in wavelength determination arises on account of the displacement of lines to be measured relative to standard lines. It is well known that the positions of spectral lines relative to a set of fiducial marks depend upon the exact alignment of the source with respect to the optical system used to study the spectra. On the problem of "displacement errors," Professor George Harrison made some interesting observations in his Preface to the "MIT Wavelength Tables."[1] In particular, he stated, "When successive exposures to a spectrum to be measured and to a standard spectrum are made, though

---

[1] Massachusetts Institute of Technology, Wavelength Tables between 10,000 Å and 2000 Å, Wiley, New York, 1939. [The authors express their gratefulness to the M. I. T. Press, which now holds the copyright for the M. I. T. Wavelength Tables, for permitting them to reproduce the sentences quoted here (1965).]

the photographic plate remains undisturbed between the two exposures and the two sources are apparently similarly imaged on the grating, displacements as great as 0.02 Å are sometimes found between the two spectra . . . these shifts probably arise from dissimilar illumination of the grating face, causing differences in line shape or from temperature changes of the grating . . . . " Although these remarks were made in connection with photographic observations of spectra, they are entirely applicable to studies in the infrared. Therefore, when making measurements on absorption lines, there is an obvious advantage in employing absorption standards, because the same source of continuous radiation and optical system would then be used both for the standards as well as the unknown spectra. Likewise, in the case of emission work, it would be worthwhile to produce the emission standards in the same source as that used for exciting the emission spectra to be measured.

B.   GRATING CONSTANT METHOD OF DETERMINING SPECTRAL POSITIONS

During the early years of studies with high-resolution infrared spectrographs, the method widely used for determining spectral positions consisted of measuring the angular rotation of the grating. For this purpose, the grating was mounted on a graduated circle and the grating equation mentioned in Chapter I, viz. $\nu = nK \csc \theta$, was employed in calculating the wave numbers of the spectral lines. It may be recalled that in this equation n is the spectral order, K (expressed in cm^{-1}) is the instrument constant, and θ the angle between the central image and the spectral line located at ν cm^{-1}. The constant K is a function of the groove spacing of the grating and the particular optical arrangement of a spectrograph. It has been customary to calculate the constant K by measuring the angle θ at which standard atomic lines of helium, mercury, or some other element are observed. The value of K so established can then be used in subsequent evaluations of the wave numbers of spectral lines for which the values of θ are measured. This type of measurement does not yield very accurate results. The limitations of this technique are dependent on the accuracy with which the angle of rotation of the grating can be determined; furthermore, there is the inherent assumption that K does not change between the time the atomic standards are recorded and the time the angular positions of the infrared lines are determined.

This technique of using the constant K was popular because of the speed with which it allowed the spectral positions of infrared lines to be determined. It should be acknowledged that the method did, to some extent, make possible the identification and interpretation of the rotational structure of vibration rotation bands of several polyatomic molecules. However, there have been uncertainties with regard to the molecular constants obtained from data calibrated in this manner.

C.   MODERN GRATINGS AND HIGH-RESOLUTION INFRARED SPECTRA

During recent years, tremendous advances have been made in the art of ruling large diffraction gratings of excellent quality approaching perfection. Until comparatively recently, most high-resolution studies in the infrared made use of gratings ruled at the Johns Hopkins University, first by Dr. R. W. Wood, and subsequently by Dr. John Strong and his associates. The first of the new large gratings was produced by Babcock and Babcock[2] at Mt. Wilson. Almost simultaneously, Dr. David Richardson and his associates at Bausch and Lomb, were able to rule gratings of comparable size and perfection. In addition, the replication techniques developed at Bausch and Lomb preserved the quality and precision of the original rulings. This success in replication techniques resulted in the ready availability of high quality diffraction gratings. In the year 1960, Harrison and Stroke[3] ruled echelles of an even higher degree of perfection by making use of interferometric control of the ruling engine. Harrison and Stroke have produced rulings in excess of 10 in. in width.

Also, at the present time, we have available for use in the near infrared (1-5 μ), highly sensitive photoconductive detectors, particularly PbS and PbSe and the photovoltaic detector InSb. A summary of the characteristics of these detectors, with particular reference to PbSe, was given by Bode *et al.*[4]; and several other articles which appeared in the June 1965 issue of *Applied Optics* (Vol. 4, pp. 631-766) also dealt with the problem of detectors.

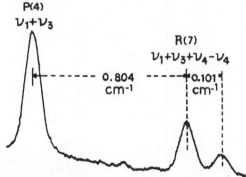

Fig. 2. The spectral interval shown is from P(4) of the $\nu_1 + \nu_3$ band of C_2H_2, and the R(7) doublet of the "hot" band $\nu_1 + \nu_3 + \nu_4 - \nu_4$ of C_2H_2.

Employing the modern gratings and the newer detectors in infrared spectrographs, it has been possible to obtain narrow spectral lines in the region 1-5 μ (see Fig. 2). It has been found that the use of a "wedge scanner" (see Section D for details) allows determination of spectral positions of sharp single lines to a very high degree of precision.

Rank[5] has discussed in detail how the modern precision obtained in the measurement of the rotational lines of vibration rotation bands of diatomic and linear triatomic molecules has led to a value for the velocity of light comparable to that determined by other methods in physics. The measurement and analysis of the rotational structure of the infrared bands of diatomic and linear polyatomic molecules lead us to an accurate value for the ground state rotational constant B for these molecules. B is essentially the reciprocal of the moment of

[2] H. D. Babock and H. W. Babcock, *J. Opt. Soc. Am.* **41**, 776 (1951).
[3] G. R. Harrison and G. W. Stroke, *J. Opt. Soc. Am.* **50**, 1153 (1960).
[4] D. E. Bode, T. H. Johnson, and B. N. McLean, *Appl. Opt.* **4**, 327 (1965).
[5] D. H. Rank, *J. Mol. Spectry.* **17**, 50 (1965).

inertia of the molecule $B = h/[8\pi^2 (cI)]$. The same constant B can be determined in pure frequency units by means of studies made with microwave techniques. The ratio $B_{microwave}/\ B_{infrared} = c$, the velocity of light. For example, from the data for HCl^{35}, a value of 299 792.8 ± 0.4 km/sec has been obtained for c.

D.   USE OF A "WEDGE SCANNER" WITH AN INFRARED SPECTROGRAPH

The principle of a "wedge scanner" was first discussed by Shearer and Wiggins.[6] The location of a wedge scanner in relation to the other parts of a high-resolution Littrow-type infrared spectrograph[7] is shown in B of Fig. 3. This instrument is equipped with a 73.25 grooves/mm ruled echelle indicated as G in Fig. 3. The grating master was ruled by Professor George Harrison at MIT. It has a blaze angle of 63°25'. The replica used in the

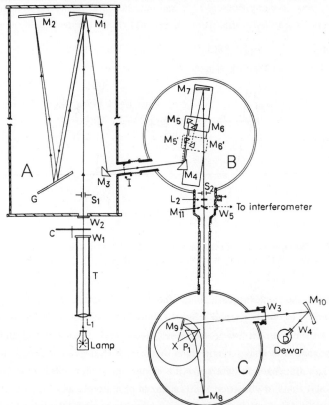

Fig. 3. A high-resolution infrared spectrograph. A shows the primary monochromator; M_1 is a spherical mirror of 5 m focal length; M_2 is the plane mirror used to double pass the grating; the spectrum comes to a focus at the position indicated by I; B is the wedge scanner; L_2 is a fluorite field lens, and M_{11} is an auxiliary mirror which can be inserted into the beam as required; C shows the order sorter, and D is the detector; the mirror M_{10} is an off-axis parabola.

[6] J. N. Shearer and T. A. Wiggins, *J. Opt. Soc. Am.* **45**, 133 (1955).
[7] D. H. Rank, *Rev. Mod. Phys.* **34**, 577 (1962).

spectrograph was prepared from the master by Bausch and Lomb. In Fig. 3, S_1 is the entrance slit of the spectrograph, M_1 is a spherical mirror of 5 m focal length which collimates the beam entering the slit S_1, and M_2 is a plane mirror used to double pass the grating. The spectrum comes to a focus at the position indicated by I. M_3, M_4, M_5, and M_6 which appear to be prisms are actually mirrors. The faces of prisms are usually flat to a high degree of perfection and, therefore, have been aluminized and employed in the wedge scanner section in place of plane mirrors. M_7 is a spherical mirror that transfers the spectrum imaged at I to exit slit S_2. The "order sorter" (commonly known as the "foreprism monochromator") is shown at C, and the infrared detector at D.

The operating principle of the wedge scanner is based on the scanning device used in a coincidence range finder. In this instrument, a weak prism or optical wedge is placed in one of the convergent beams of the range finder. The displacement of the image is proportional to the distance of the wedge from the focus. In this way, the displacement of one of the images can be smoothly and continuously varied by translating the wedge toward and away from the focus. In the wedge scanner the optical wedge has been replaced by a pair of mirrors (M_5 and M_6) arranged so as to slighly deviate the line of sight. The mirror system has the advantage of being achromatic. The deviation can be set at any convenient small angle. A typical angle of deviation is $1.5°$.

The grating is kept fixed at a particular position and the wedge scanner is used to scan the spectrum imaged at I. This is accomplished by translating the pair of mirrors M_5 and M_6 in the manner indicated in Fig. 3. The wedge is shown in two alternate positions. The mirror M_4 reflects the spectrum imaged at I onto the wedge M_5 and M_6 so that the direction from M_4 to M_5 is parallel to the movement of the scanner. A photograph of the scanner itself is shown in Fig. 4. A region of the spectrum approximately 1 cm in extent at I is scanned. This corresponds to an extent of about 50 Å in the first order of a 73.25 grooves/mm grating blazed for an angle of $63°25'$. Therefore, when a wedge scanner is used, it is important to have a standard line available for every 50 Å interval. Figure 5 shows the regions in which the absorption standards listed in Chapter III can be observed in higher spectral orders when a 73.25 grooves/mm echelle blazed for an angle of $63°25'$ is used. These absorption standards are so located that for spectra occurring in the region 1-5 μ, it is possible to use a coarse grating like the one with 73.25 grooves/mm and measure any unknown wavelength with reference to a standard by using the wedge scanner. In order to use this method effectively, it is important first to know the spectral position to an accuracy of about ± 0.1 cm^{-1}. This permits the selection of an appropriate standard of either CO, HCN, or N_2O. In addition to the gas under study, the requisite amount of CO, HCN, or N_2O should be introduced into the optical path of the spectrograph. While the wedge scanner traverses from one end to the other of the screw to which it is attached, the infrared line for which the wave number is desired and the absorption standard of HCN, CO, or N_2O are recorded. The distance r on the chart paper between the infrared line and the infrared standard can be measured. If the dispersion $d\nu/dr$ is known (see the next section), it should be possible to evaluate the position of the infrared line relative to the standard. In order to facilitate the

Fig. 4. A photograph of the wedge scanner represented in B of Fig. 3. When this photograph was taken, a slit was located at I in Fig. 3; the flat mirror M_4 shown in B does not appear in the photograph.

use of the infrared absorption standards of CO, HCN, or N_2O, the first order wave numbers for each of the absorption standards which actually are observed in a higher spectral order are furnished in Appendix III. It may be noted from the information presented in Fig. 5, that the first order wave numbers refer to a 73.25 grooves/mm echelle blazed for 63°25'. However, the wave numbers given can also be used with other gratings having similar groove spacings.

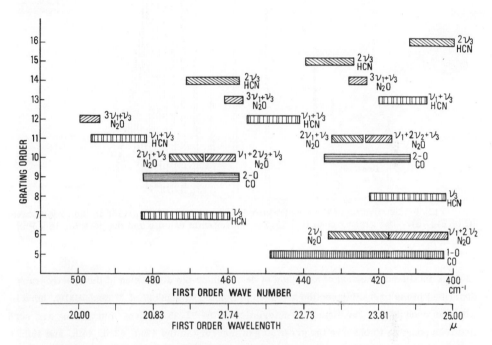

Fig. 5. Grating orders in which some of the vibration rotation bands of the CO, N_2O, and HCN molecules can be observed with a 73.25 grooves/mm echelle.

When a coarse grating is used in higher spectral orders, the foreprism monochromator should be able to separate the grating orders. Figure 6 demonstrates that this can be easily accomplished by employing a LiF prism. In the spectrograph described above (Fig. 3) the LiF prism was installed after the exit slit of the grating section. This was done to avoid possible "displacement errors" arising from different illuminations of the grating in different orders. The narrow slit widths used for the grating spectrograph were responsible for this complete separation of orders.

It has also been found[8] that a large 60° NaCl prism (17.8 cm base and 14 cm height) with a standard Littrow-type mirror behind it for scanning the spectrum will separate these grating orders satisfactorily.

[8] K. Narahari Rao and H. H. Nielsen, *Appl. Opt.* **2**, 1123 (1963).

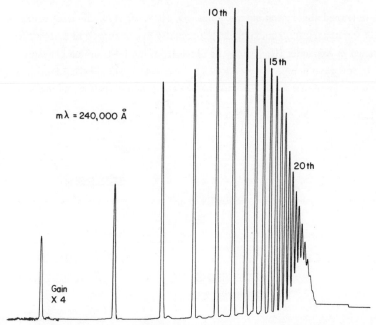

Fig. 6.   Separation of the spectral orders employing a LiF prism as the order sorter (C of Fig. 3).   The grating used is a 73.25 grooves/mm echelle and the detector is a PbS cell.

Determination of Dispersion $d\nu/dr$ versus ν:

A detailed discussion of a method of determining the dispersion of the spectrograph described in the preceding section has been presented by Rank *et al.* [9] Essentially, here is what has been done.   The blaze characteristics of the 73.25 grooves/mm grating are such that it is possible to observe the green line of mercury in the 42nd, 43rd, 44th, and 45th orders in the angular stretch where the grating is useful.   The broadened green line from a high pressure mercury arc was employed as a source of light, and light from the exit slit of the spectrograph (see Fig. 3 marked "to interferometer") was made parallel by means of a concave mirror and allowed to fall on an evacuated Fabry-Perot etalon with a known spacer of $t = 13.193$ mm.   The broadened green line was caused to scan over the exit slit by means of the wedge scanner.   The intensity variation produced in a small part of the central ring of the interference pattern was detected by a photomultiplier, the output of which was recorded.   The separation of the peaks in the "fringes" appearing on the record is equal to $1/2t$ cm^{-1}.   The values of $d\nu/dr$ have been measured for each of the grating orders viz. 42nd, 43rd, 44th, and 45th.   This information allows one to plot $d\nu/dr$ versus the first order wave numbers corresponding to the 42nd, 43rd, 44th, and 45th orders of the green line of mercury, and the curve can be used for determining $d\nu/dr$ at any ν.   If $d\nu/dr$ is known for the first order wave number, the value of $d\nu/dr$ for the mth order is simply $[1/m(d\nu/dr)_{\text{first order}}]$.

[9] D. H. Rank, D. P. Eastman, W. B. Birtley, G. Skorinko, and T. A. Wiggins, *J. Opt. Soc. Am.* **50**, 821 (1960).

E.   COARSE ECHELLES

For the discussion presented in this monograph, three different coarse gratings have been chosen as examples. There is no implication, however, that the techniques are limited to gratings with this type of groove spacing. In the preceding section, it has been noted that the available absorption standards of CO, HCN, and N_2O are adequate for spectrographs equipped with a 73.25 grooves/mm echelle blazed at an angle of 63°25'. The other two coarse echelles currently used[10] with high-resolution infrared spectrographs are the Bausch and Lomb replicas with 40 and 30 grooves/mm. The 40 grooves/mm Bausch and Lomb plane grating has its primary blaze at 23°, at which angle the 19 μ radiation can be observed in the first order. This grating can be employed at its secondary blaze of 67° for use in

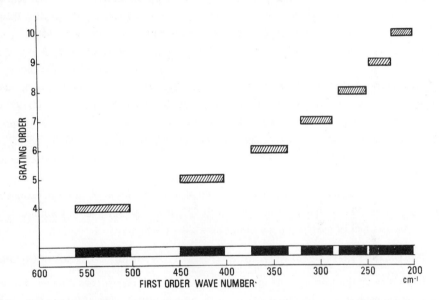

Fig. 7. Spectral orders in which the 1-0 vibration rotation band of the CO molecule can be observed with a 40 grooves/mm Bausch and Lomb plane replica grating (primary blaze, 23°), used "echelle fashion."

higher spectral orders between incident angles of 50° to 80°. Figure 7 shows the regions in which the 1-0 vibration rotation band of the CO molecule can be observed with the 40 grooves/mm grating. It seems possible that rotational lines of this band can be used as standards in most of the angular stretches in which the grating can be employed for measuring infrared spectra. The mixing of the orders is even better with a 30 grooves/mm grating.[11] In view of this advantage, the first order wave numbers of the rotational lines of the 1-0 CO band actually observed in higher spectral orders are also furnished in Appendix III.

[10] K. Narahari Rao, W. W. Brim, V. L. Sinnett, and R. H. Wilson, *J. Opt. Soc. Am.* 52, 862 (1962).
[11] T. K. McCubbin, Jr., J. A. Lowenthal, and H. R. Gordon, *Appl. Opt.* 4, 711 (1965).

In this monograph, the word "echelle" has been employed to describe plane gratings with coarse groove spacings used at incident angles of about 60° in high spectral orders. There are several advantages in using echelles in high-resolution infrared spectrographs, especially when the problem is one of determining spectral positions. In the preceding paragraph, it has been noted that it is relatively easy to locate standard lines in one of the higher grating orders with an echelle.

The second important advantage in using echelles lies in the possibility of obtaining better angular dispersion. An example will be given to clarify this point. Referring to the grating equation $\nu = nK \csc \theta$, one can immediately deduce $|d\nu/d\theta| = \nu \cot \theta$. Suppose we are interested in studying spectra occurring in the region of 10 μ (1000 cm^{-1}). When observations are made with a grating at $\theta = 26°35'$ (cot $\theta \approx 2$), an interval of 1 reciprocal cm ($\Delta\nu = 1$ cm^{-1}) at this wavelength (i. e., 10 μ) can be scanned by rotating the grating through 1.72 min of arc; however, when observations are made with an echelle at $\theta = 63°25'$ (cot $\theta \approx 1/2$), the grating must be rotated through a larger angular interval (about 6.88 min of arc) to observe the same 1 cm^{-1} interval. Therefore, in observations made with an echelle, the wavelength error caused by a given error in angular measurement is much less pronounced. When spectra are traced by a recorder, the advantage in the angular dispersion ($d\nu/d\theta$) obviously appears as linear dispersion ($d\nu/dx$; x is a distance measured on the chart paper).

Finally, it is of interest to note that an echelle facilitates the use of wider physical slit widths. For instance, referring to Eq. (I-2): if observations are obtained at an angle of 26°35' with a spectrograph of focal length 4 m, a spectral slit width of 0.1 cm^{-1} at a wavelength of 10 μ corresponds to a physical slit width of 0.4 mm; however, if observations are made of the same 10 μ radiation at an angle of 63°25', the same spectral slit width of 0.1 cm^{-1} can be attained by opening up the slits to a width of 1.6 mm.

F.   RECORDERS EMPLOYED WITH HIGH-RESOLUTION INFRARED SPECTROGRAPHS

Depending on the nature of the investigational work undertaken, two different types of recorders are used in infrared laboratories at the present time. One of these is equipped with a single pen for tracing spectra and the second type has two pens. With a double-pen recorder, it is possible to trace simultaneously the spectrum to be measured as well as a reference spectrum. Whichever technique is convenient, it is now possible to devise ways and means of obtaining good wavelength measurements in the infrared.

Use of a Single-Pen Recorder:

When a wedge scanner is employed for the observation and measurement of infrared lines (see Section D for details), the use of single-pen recorders is adequate. In a particular scan, a standard line is always recorded along with the spectral lines requiring wavelength measurement. Rank and his associates note that it is possible to obtain accuracy of measurements to ±0.001 cm^{-1} with a wedge scanner. Therefore, precision seems to be dictated solely by the line width and signal-to-noise ratio. Since this technique permits only the scanning of rather narrow spectral regions at one time, it seems ideal for use when high precision in the wavelength measurement of single lines is desired.

It is often convenient to scan spectra by rotating the grating. If a satisfactory drive mechanism is used for rotating the grating (see pp. 165-168 for details of some of the systems currently used), an accuracy of ±0.005 cm^{-1} in the determination of spectral positions is not at all difficult to achieve. The procedure that appears to be satisfactory when single-pen recorders are used will be described first. Following this, the techniques pursued by laboratories using double-pen recorders will be elaborated.

McCubbin[12] has been able to obtain precise measurements for the rotational lines of linear polyatomic molecules like N_2O and CO_2 by adopting a procedure well-known to optical spectroscopists. The technique is easily understood with the aid of Fig. 8. With the use of an 80 grooves/mm grating, it is possible to observe the 1-0 band of CO in the third order of the grating. First, the CO lines are recorded on chart A [for example, from R(6) through R(19)], and then the infrared spectrum to be measured is traced on chart B which includes

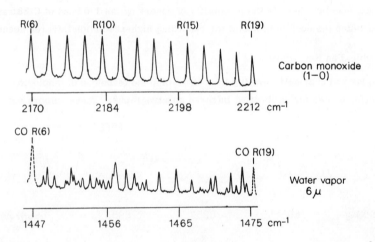

Fig. 8. A technique that illustrates the use of single-pen recorders for determining precise spectral positions relative to the rotational lines (absorption standards) of the 1-0 band of the CO molecule. The grating employed for recording the spectra shown here is a Bausch and Lomb plane replica with 80 grooves/mm.

two CO lines, one at the beginning [R(6) in Fig. 8] and another at the end [R(19)]. If the distance between these two CO lines [R(6) and R(19)] is maintained at some preset limit in each chart, the following procedure is satisfactory for evaluating the wave numbers of the unknown lines. The distance of each of the CO lines [R(7) through R(19)] is measured from R(6). If linear dispersion is assumed between lines R(6) and R(19) it is easy to derive the constants dv/dx and v_0 in the equation

$$v = (dv/dx) X + v_0 \qquad\qquad (V-1)$$

where X = distance from line R(6) on chart A. These constants are then used to calculate

[12] T. K. McCubbin, Professor of Physics, Pennsylvania State University, University Park, Pennsylvania (private communication, 1965).

the wave numbers of the CO lines R(7) through R(18) and the calculated values can then be compared with the standard positions listed in Chapter III. The differences "calculated-observed" may then be plotted against X to give the "correction curve." Then, all the lines in the unknown spectrum are measured from the line R(6). By using Eq. (V-1) and the correction curve, the wave numbers of all the lines in chart B can be calculated.

In this technique, it is important that linear separations on the chart paper between standards like R(6) and R(19) be maintained on both charts. Rao et al. [13] have reported that it is possible to record the unknown spectra between CO standards without stopping the scanning of the spectra by using a suitable manipulation of the foreprism monochromator. In this way, Eq. (V-1) can be employed even more effectively.

The data obtained with an 80 grooves/mm grating were mentioned in discussing the above procedure in view of the ready availability of the illustration shown in Fig. 8; the technique is more suitable, however, for a coarser grating. For instance, with a 30 grooves/mm echelle, there is the possibility of observing the 1-0 band of CO at any angular position at which the echelle is useful for recording high-resolution infrared spectra.

Use of Double-Pen Recorders:

Douglas and Sharma[14] were the first to develop the technique of employing a Fabry-Perot interferometer for calibrating infrared spectrographs. These authors sent two beams

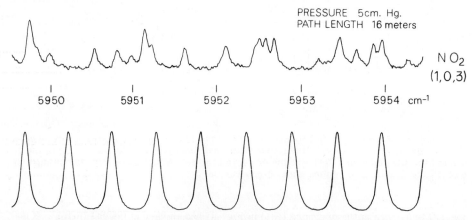

Fig. 9a. Infrared absorption spectra and third order white light fringes recorded simultaneously on the two pens of a double-pen recorder. The grating used was a Bausch and Lomb plane replica with 15,000 lines per inch. Fringe spacing was determined to be 0.535811 ± 0.000001 cm^{-1}. The source of visible light for the "fringes" was a 100 watt zirconium arc and the detector an RCA 1P21 phtomultiplier. The etalon has fused quartz plates aluminized to about 80% reflectivity and a 3 mm fused quartz spacer. The etalon is followed by a constant-deviation Wadsworth prism-mirror arrangement to separate orders of visible light coming from the monochromator. The Wadsworth mount may be rotated to select the desired order of visible light and to maximize the response of the photomultiplier. All elements except the source and detector are in the vacuum chamber of the spectrograph (courtesy of Dr. T. Harvey Edwards).

[13] K. Narahari Rao, L. R. Ryan, and H. H. Nielsen, J. Opt. Soc. Am. 49, 216 (1959).
[14] A. E. Douglas and D. Sharma, J. Chem. Phys. 21, 448 (1953).

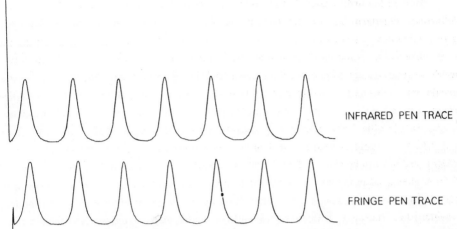

INFRARED PEN TRACE

FRINGE PEN TRACE

Fig. 9b. Dynamic pen separation for the two pens of a double-pen recorder obtained by sending the same fringe signal to the two pens. A normal spectral record consists of successive sections of a dynamic pen separation segment, a segment of absorption standard lines, the absorption spectrum to be studied, a further segment of standard lines, and a final dynamic pen segment. No stops or changes in driving speed of the grating or chart should be made from start to finish. It is important to obtain the dynamic pen separation because the separation of the pens is not the same for a moving chart as for a stationary chart (courtesy of Dr. T. Harvey Edwards).

of radiation through the grating optics at the same time. In one instance, the continuous infrared radiation passes through an absorption cell filled with the gas to be studied. After passing through a monochromator and a grating spectrograph, it reaches an infrared detector. The other beam traverses an etalon with a suitable spacer. Fringes (Edser-Butler bands or channeled spectra) of visible light in the region around 5000 Å are detected with a photomultiplier in the higher orders of the grating. One of the two pens of a double-pen recorder records the fringe system, and a recording of the infrared absorption spectrum is simultaneously obtained with the second pen. These fringes serve to impress a "wave number scale" on the infrared spectrum to be measured. The fact that the fringes are equally spaced in wave number units offers some computational advantages. The spacings of the fringes and their actual positions can be determined by interpolation between suitable standard absorption lines, recorded as part of the infrared trace (see Fig. 9b).

Employing essentially the same method, Plyler and his co-workers[15] published measurements for the rotational lines of several vibration rotation bands of polyatomic molecules occurring in the near infrared. Also, at the University of Lund, Lund, Sweden,[16] the above procedure was used for determining spectral positions of atomic lines observed in the near infrared reached by a PbS detector. The illustration and caption in Fig. 9 outline the technique followed by Professors C. D. Hause and T. Harvey Edwards at the Michigan State University, East Lansing, Michigan. These authors also consider the method entirely satisfactory for investigations in the near infrared of 1-5 μ.

[15] E. K. Plyler and E. D. Tidwell, *Mem. Soc. Roy. Sci. Liege* [4] **18**, 426 (1957)
[16] See for example, U. Litzen, *Mem. Soc. Roy. Sci. Liege* [5] **9**, 29-36 (1964).

Some of the precautions that should be observed in employing this technique are the following. Standard lines for determining wave numbers of fringes must be recorded during a run; otherwise, fringes cannot be used. Regulation of the temperature of the etalon plates is important. A limitation to the use of this method is the requirement that the spectral slit width be small enough to prevent the fringes from being washed out. In practice, the width should not exceed half the spacing of the fringes. Furthermore, the spacer thickness becomes a problem when fringes formed in the visible are to be used for measurements at longer wavelengths. For instance, if measurements in the region of 10 μ are to be made and if it is desirable to obtain fringes of 0.5 cm^{-1} separation, then the thickness of the spacer will have to be about 0.5 mm. These specifications apply to the fringes formed at 5000 Å when used in the 20th order of the grating. The problem of obtaining closely spaced visible light fringes presents more difficulties when measurements are made at even longer wavelengths. However, if infrared transmitting etalon plates are used, and fringes are produced in the near infrared region of the spectrum accessible to the photoconductive detectors, this difficulty can probably be minimized. The problems connected with the use of a Fabry-Perot interferometer are not insurmountable, but one may not enjoy encountering them in day-to-day work.

Use of Atomic Lines As Wave Number Markers:

Instead of using the interferometer fringes, Rao *et al.* [17] employed the atomic lines of neon excited in a hollow cathode discharge tube as wave number markers (see Fig. 10). The use of neon lines was most satisfactory when Bausch and Lomb plane replica gratings with 80 grooves/mm or 40 grooves/mm were utilized. It should again be emphasized that these specific gratings are mentioned here because the experiments by Rao *et al.* were performed with them. However, the techniques should be applicable with other gratings having similar groove spacings and blaze characteristics. It was noted that the neon lines occurring in the visible could be observed in several grating orders over the wide angular stretches at which these gratings could be used for stuying high-resolution infrared spectra; therefore, it was possible to obtain an adequate set of reference points to allow the determination of the wave number scale. As in the case of the interferometer technique, the absolute positions on this wave number scale should be fixed by recording suitable infrared standards on the pen employed for recording the infrared spectra under study. It is important to remember that the same optical paths in the spectrograph are involved when the infrared standards are recorded and also when the infrared spectra to be measured are observed.

It should be mentioned that it is not satisfactory to use the neon spectrum with gratings having smaller groove spacings (for instance, 300 grooves/mm or 600 grooves/mm). With such gratings, the visible neon lines are observed in a relatively few grating orders and, therefore, one would not have a sufficient number of points to determine the wave num-

[17] K. Narahari Rao, T. J. Coburn, J. S. Garing, K. Rossmann, and H. H. Nielsen, *J. Opt. Soc. Am.* 49, 221 (1959).

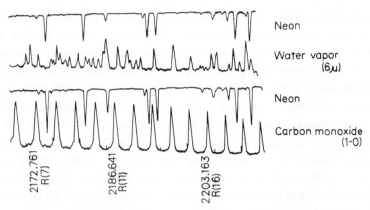

Neon

Water vapor
(6μ)

Neon

Carbon monoxide
(1-0)

2172.761
R(7)

2186.641
R(11)

2203.163
R(16)

Fig. 10. A technique employed to illustrate the use of neon lines as "wave number markers," and the rotational lines of the 1-0 band of CO as absorption standards. The grating employed for recording the spectra shown here is a Bausch and Lomb replica with 80 grooves/mm.

ber scale. However, it seems possible[18] to use the atomic lines of thorium with these gratings. A thorium lamp excited in a resonant cavity by a 2450 Mc "microtherm" oscillator shows an intense thorium spectrum. The number of thorium lines occurring in the visible is so large that it is necessary to use appropriate filters to avoid blending of the spectral lines occurring in different grating orders. It has been the experience in several laboratories that the "microtherm" oscillator causes electronic interference when a thermocouple is used to detect the infrared radiation. In such a case, it may be advantageous to use the hollow cathode discharge tube as a source.

Coarse Echelles and Double-Pen Recorders
(Use of Absorption Lines As Wave Number Markers):

It has been noted in an earlier section of this chapter that large ruled echelles with 10×5 in. ruled surfaces are now available. The following comments are made with reference to an echelle with 30 grooves/mm blazed at an angle of 63°25', because the prospects of producing even larger echelles of this type seem to be good. It has been observed that the characteristics of this echelle are such that outside of a small angular stretch near the blaze region, it is not possible to observe the atomic lines occurring in the visible. The difficulties will be even greater if interferometer fringes formed in the visible are to be recorded with a grating having such blaze characteristics and groove spacings. Since the 1-0 band of CO can be observed at any angular stretch where the grating is useful, the infrared lines to be measured can be interspersed between the rotational lines of this fundamental vibration rotation band of the CO molecule by suitable manipulation of the foreprism monochromator. In this manner, the unknown lines can be measured relative to the CO standards. However, this procedure is not always satisfactory, because the molecule under in-

[18] D. Copen, M. S. Thesis, The Ohio State University (1964).

vestigation may have strong absorption bands in the region where the 1-0 band of the CO molecule occurs. Furthermore, in the case of complex spectra, it is more satisfactory to be able to obtain a continuous record of the complete spectrum. Therefore, it would be of interest to devise other means of producing a wave number scale by employing the second pen of a double-pen recorder. At the time this monograph was being written, a solution to this problem was indicated in the work of Rao *et al.*[19] at The Ohio State University. Instead of visible interferometer fringes or the visible atomic lines, the second pen was used to record the absorption spectrum of either the 1-0 or the 2-0 band of the CO molecule. In other words, the second beam passing through the spectrograph also comprises continuous infrared radiation (from a Nernst glower, a zirconium arc lamp, or a carbon furnace,[20] depending on the intensity requirements). It was possible to resolve the higher grating orders in the 1-3 μ region by employing a foreprism monochromator having adequate dispersion and a PbS cell for detecting the infrared radiation. In the particular arrangement used at The Ohio State University, the second beam illuminated only a 4 or 5 mm length in the

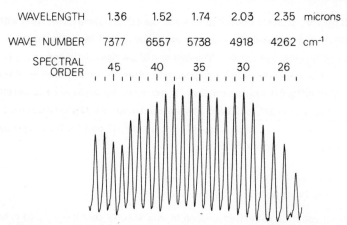

WAVELENGTH	1.36	1.52	1.74	2.03	2.35 microns
WAVE NUMBER	7377	6557	5738	4918	4262 cm⁻¹
SPECTRAL ORDER	45	40	35	30	26

Fig. 11. Separation of the grating orders to allow use of the 2-0 band of CO observed in the 26th order as "wave number markers." Other suitable absorption bands of HCN, for instance, can be utilized for the same purpose. The foreprism monochromator used is a Beckman IR-3 spectrometer equipped with two double-passed 60° NaCl prisms; the source is a Nernst glower, and the detector an uncooled PbS cell; the grating employed is a 30 grooves/mm, 10 × 5 in. ruled echelle; the grating master was ruled at MIT, and the replica prepared at Bausch and Lomb.

entrance slit of a 4 m focal length Littrow-type spectrograph, whereas the regular infrared beam illuminated a length of about 1½ in. Figure 11 shows that the higher grating orders are well resolved, permitting observation of the 2-0 band of CO in the 26th grating order. The rotational lines of this 2-0 band serve to produce the wave number scale on the infrared spectrum to be measured. The 1-0 CO band recorded on the same pen employed for tracing

[19] K. Narahari Rao and associates, The Ohio State University (1965); see A. Karger, M. S. Thesis, The Ohio State University (1965).

[20] R. N. Spanbauer, P. E. Fraley, and K. Narahari Rao, *Appl. Opt.* 2, 1340 (1963).

the unknown infrared spectrum is still used to determine the absolute positions on this wave number scale. Figure 12 illustrates the use of a 60° NaCl prism (with a standard Littrow mirror behind for scanning the spectra) for resolving the grating orders so that the 1-0 band of CO can easily be observed in the 13th order of the grating even when a Nernst glower is used as a source of continuous infrared radiation and a thermocouple employed as a detector. It was also noticed that the blaze characteristics of the 30 grooves/mm echelle are such that the 2-0 CO band could not be observed in the entire angular stretch at which the grating is useful for investigations at longer wavelengths between 6 and 15 μ. In such an event, it is possible to use the 1-0 CO band to produce a wave number scale utilizing the second pen of the double-pen recorder. For observing this 1-0 CO band one obviously has to employ a detector such as PbSe or InSb. In such a system, the wave number scale as well as the absolute infrared positions on this scale are obtained by utilizing the

WAVELENGTH	4.69	6.10	7.62	8.71	10.17	12.20	microns
WAVE NUMBER	2132	1639	1312	1148	983	820	cm⁻¹
SPECTRAL ORDER	13	10	8	7	6	5	

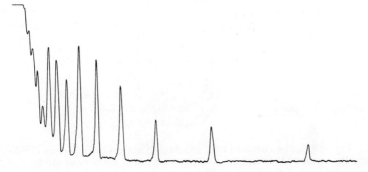

Fig. 12. Separation of the spectral orders employing a 60° NaCl prism with a standard Littrow mirror behind it to scan the spectra. The particular prism used here has a 17.8 cm base and is 14 cm in height. The grating is a 30 grooves/mm echelle; a Nernst glower is used as a source of continuous infrared radiation, and the detector employed is a thermocouple. At the grating angle where this scan of spectral orders was taken, the 1-0 CO band would be observed in the 13th order of the grating. (The trace shown here was obtained by Mr. K. L. Yin at the Ohio State University.)

same fundamental vibration rotation band lines of the CO molecule. It is indeed true that one can employ any other suitable absorption lines for impressing the wave number scale. However, in view of the simplicity of the structure of the CO bands it is more advantageous to use CO lines for this purpose.

Grating Drive:

Since the mechanism employed to rotate the grating plays a significant role in the accuracy which can be obtained for determining spectral positions in the infrared, this section will be devoted to a description of two different types of grating drives. One of these

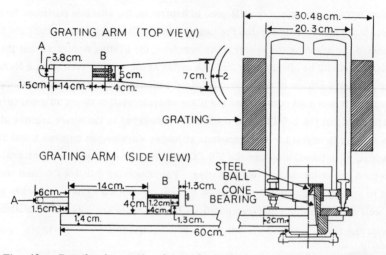

Fig. 13a. Details of a grating drive which employs a precision screw to guide the rotation of the grating.

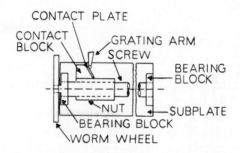

Fig. 13b. The contact block attached to the precision screw. When the screw is rotated, the contact block and the nut assembly move parallel to the axis of the screw.

devices (purchased from David Mann and Co., Lincoln, Massachusetts) has been installed in the high-resolution infrared spectrographs [21, 22] at The Ohio State University. Essentially, it consists of a precision screw employed to guide the rotation of the grating. A lever arm about 60 cm long is connected to the mounting on which the grating holder is supported. A ball (A in Figs. 13 and 14) has been fastened to the end of this lever arm and allowed to rest on an optically flat quartz plate attached to the contact block. The plate and the contact block are part of a nut assembly which moves parallel to the axis of the screw when the screw is rotated. Further details of this grating drive are shown in Figs. 13 and 14.

[21] See Rao and Nielsen [8].
[22] J. G. Williamson, M. S. Thesis, The Ohio State University (1964).

A second type of grating drive mechanism, described by McCubbin and his co-workers[23], is shown in Fig. 15. The drive is so designed that coarse rotation for positioning or for rapid scanning is easily obtained. The bearing consists of three concentric cones. The largest cone is a flanged iron casting having a fixed position with respect to the chassis of the instrument. An intermediate steel cone equipped with a lever arm runs inside the large cone. This arm, 37 cm long, is moved by a tangential push rod and 25 mm micrometer screw and provides about 10° of very precise and smooth rotation. The end of the rod is steadied by means of a ball race which rolls upon a track. A long steel spring of the kind used as a drive belt in motion picture projectors supplies the force that keeps the steel pusher plate on the arm in contact with the ball on the push rod.

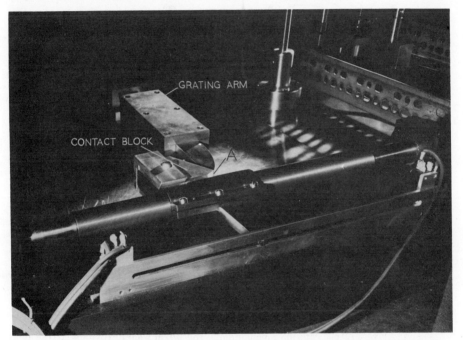

Fig. 14. A photograph of the precision screw represented in Figs. 13a and 13b. The picture also shows part of the grating arm, the contact block, and the nut assembly.

The inner cone, upon which the grating holder is mounted, is turned by a worm wheel of about 15 cm diam. The worm drive is used to set the position of the grating with respect to the arm and tangent screw drive and to make rapid scans of the spectrum. The worm and its shaft turn on bearings which are mounted on the arm. The worm, which moves several millimeters as the arm travels, is turned by means of a telescoping shaft and universal joints which are not shown in the figure.

[23] See McCubbin *et al.*[11]

The cones have half angles of $5\frac{1}{3}$° and the diameters at the small ends are 2.5 cm and 6.4 cm. The large stationary cast iron cone is 8.4 cm long. A single 8 mm steel ball on the axis which bears upon the hardened polished end of a steel adjustment screw supports the weight of the moving parts and prevents excessive friction.

The cones were lapped by hand after they had been machined using a lathe and taper attachment. Several hours of hand lapping were sufficient for finishing both tapers. A very light oil was used for lapping and a fine 9 μ grade emery for finishing. This was followed by work with only the oil and no abrasive.

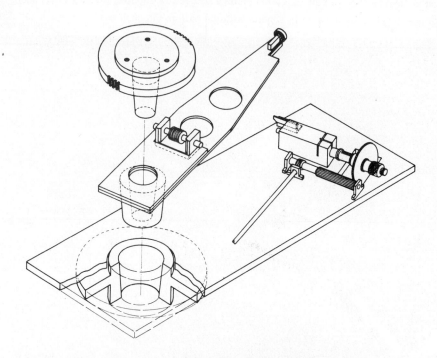

Fig. 15. The grating rotation mechanism. The worm wheel provides coarse motion, and the tangent screw and arm provide precise smooth motion for wavelength measurement. Not shown are the universal joints and telescoping shaft which drive the worm, or the single steel ball bearing which supports most of the weight of the moving parts.

Measurement of Records of Infrared and Calibration Spectra:

Accurate measurement of linear distances on the chart paper on which the spectra are recorded is another important requirement in the determination of precise values for spectral positions in the infrared. Examples will be given to illustrate some of the procedures employed for measuring chart distances. First, a measuring device based on the vernier principle will be described, and then a more sophisticated technique which uses a digitized film reader will be presented.

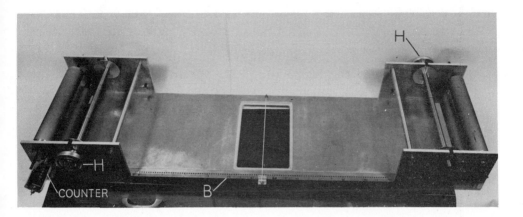

Fig. 16a. A photograph of the vernier device used to measure spectra. This figure shows the device without the chart paper in it. The handle marked H at the right of the figure had been reversed when this photograph was taken, and the correct position is shown in Fig. 16b.

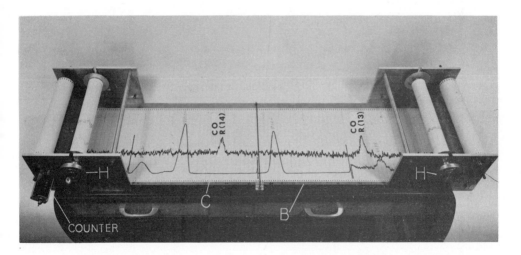

Fig. 16b. A photograph of the same vernier device shown in Fig. 16a, except that the chart paper having the spectral lines to be measured has been inserted into position.

The vernier device discussed here has been developed and constructed at The Ohio State University with the assistance of P. E. Fraley and Jeffrey Luce; photographs of this device are shown in Fig. 16. Evenly spaced holes, corresponding in size to the holes appearing on the edge of the chart paper, were drilled into a vernier bar. (This vernier bar is shown more clearly in Fig. 16a.) Figure 16 also shows the rollers and mounts for passing the record over the vernier bar. The separation x between the holes on the bar has been made such that the distance over N holes on the bar, Nx, is equal to the distance over $N+1$ holes in the chart paper, $(N+1)\bar{x}$, where $\bar{x}$ is the distance between the successive holes on the chart paper. In other words

$$Nx = (N+1)\bar{x} \tag{V-2}$$

If the chart paper is moved from left to right by turning the handles (H in Fig. 16) attached to the rollers, and a reference hole chosen on the vernier bar, then the number of holes, I, passing the reference hole toward the right may be counted with a digital counter attached to the rollers. Passage of I holes means that a distance of $I\bar{x}$ has been covered on the chart paper. The fraction of $\bar{x}$ past the reference hole (indicated by d in Fig. 17) must now be added to $I\bar{x}$, and this fraction can be obtained by noting which holes (h' and h) on the chart

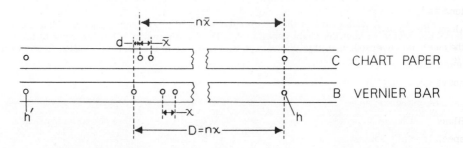

Fig. 17. A sketch of some of the sections of the chart paper and the vernier bar of the measuring device shown in Fig. 16.

best coincide with those on the bar. It should not be assumed that h' and h will always be oriented relative to the reference hole in the manner shown in Fig. 17. The number of holes in the bar between h' and h is N. If the holes in the vernier bar are numbered from left to right and the hole designated by h is n holes on the bar from the reference hole, then the distance D from the reference hole to the hole h is given by

$$D = nx \tag{V-3}$$

It follows from Eq. (V-2) that

$$D = n(N+1)(\bar{x}/N) = n\bar{x} + (n/N)\bar{x} \tag{V-4}$$

Therefore, $$d = (n/N)\bar{x} \tag{V-5}$$

Measurements of the spectral lines are usually made from some arbitarily chosen point at one end of the chart. The total distance on the chart paper from this point is given by $I\bar{x} + d$. When $n = N$, I will have increased by one unit as measured by the counter. It should be mentioned that N may vary slightly from chart to chart or even within the same chart and, therefore, it is important to keep careful records of its value. The above procedure is satisfactory because, in principle, the recorder measures the number of holes per unit time and not the distance per unit time. With $N \approx 100$ and $\bar{x} = \frac{1}{4}$ in., the standard deviation of centering on a nonblended absorption line is found to be about $3\bar{x}/100$ or about 0.2 mm. With the particular grating and chart speeds employed[24] at the 2.7 μ region, this uncertainty amounts to less than 0.002 cm^{-1}. When records of complex spectra covering large angular stretches are measured, it is convenient to use this device rather than a ruler.

An alternate method of measuring the linear positions of spectral lines on the chart paper has been successfully employed by James L. Griggs, Jr. at The Ohio State University. In this technique the measurements are made with reference to the printed lines of the chart paper rather than the punched holes since the shape of the holes sometimes becomes distorted through use.

A vertical line and a vernier scale are scribed on the underside of a transparent plexiglass overlay of 3/16-inch thickness (see Fig. 18). The vertical line should be as long as the printed lines, whereas the vernier scale can be only about an inch in length along the top or bottom of the chart paper. The scribed lines can be seen more clearly if they are drawn with a dark-colored ink. The digital counter used for counting chart paper holes can be used for counting printed lines on the chart paper if an appropriate change of gears attached to the rollers is made.

A more sophisticated measurement technique has been employed by Edwards and Blass.[25] Charts showing the infrared spectral lines and fringes are photographed using a specially-built copy stand with a permanently built-in Nikon F camera with 55 mm, $f/3.5$ Micro Nikor lens, selected because of its relative freedom from distortion. Each frame covers about 15 in. of chart (see pp. 160-161 for sample records of Edwards $et\ al.$).

The developed film is displayed on a Hydel Model 200 [Raymond Atchley, Inc., (formerly Hydel, Inc.), 2339 Cotner Ave., Los Angeles, California. Similar devices are marketed by Gerber Scientific Instruments Co., 90 Spruce St., Hartford, Connecticut, and others] semi-automatic film reader such as is used in the analysis of bubble chamber photographs, etc. The digitized output is connected to an IBM 526 summary card punch which records on computer cards the x and y coordinates of positions of interest on the film. The computer then determines the positions of the lines of both standard and sample gases relative to the fringes, and corrects for the dynamic pen separation (throughout the entire record), and for the rotation of each photographic frame due to positioning of the chart on the copy stand and the film on the projection stage. A calibration equation for fringe numbers versus the known frequencies of standard lines is obtained by a linear least squares fit, and subsequently is used to determine the frequencies of the desired spectral lines. This method of reducing the data has proved to be convenient, fast, and accurate. Reproducibility of better than 0.1 mm on the original chart is achieved, which corresponds to reproducibility in the measuring process of 0.003 to 0.001 cm^{-1} for records run with 1 cm^{-1} covering from 30 to 100 mm of chart.

[24] P. E. Fraley and K. Narahari Rao, $J.\ Opt.\ Soc.\ Am.$ **55**, 1091 (1965).

[25] Measurement of Infrared Absorption Line Frequencies Using a Digitized Film Reader, T. H. Edwards and W. E. Blass, $J.\ Opt.\ Soc.\ Am.$ **52**, 1311 (1962) (abstract).

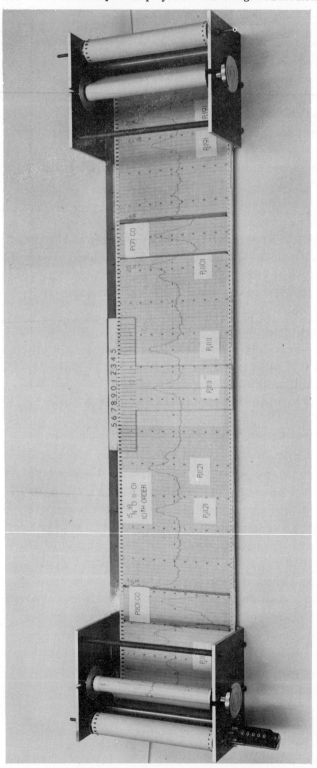

Fig. 18a. A photograph of the vernier device used in measuring spectral lines making use of the printed lines on the chart paper. The data shown are the rotational lines of the $1 \rightarrow 0$ band of the $N^{15}O^{18}$ molecule recorded in the 10th order of a 30 grooves/mm ruled echelle. A Nernst glower has been used as a source of continuous infrared radiation, and a thermocouple as a detector. The subscript on each line indicates the subband; the numeral 1 is used for the $2_{\Pi_{1/2}} - 2_{\Pi_{1/2}}$ transition, and the numeral 2 for the $2_{\Pi_{3/2}} - 2_{\Pi_{3/2}}$ transition. The appropriate $(J + 1/2)$ values are given in parentheses. The rotational lines of the $1 \rightarrow 0$ band of the $C^{12}O^{16}$ molecule observed in the 12th order were used as standards for wavelength calibrations. The $1 \rightarrow 0$ CO band was recorded by appropriately rotating the foreprism monochromator without stopping the scanning of the spectra.

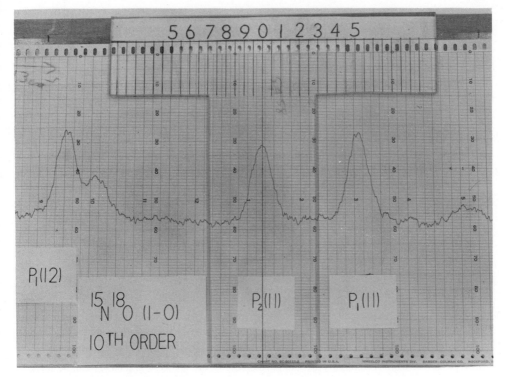

Fig. 18b. A close-up of the vernier device shown in Fig. 18a.

G.   THE USE OF COMPUTER PROGRAMS

At the present time, one likes to think in terms of "speeding-up" the processing of observational data by employing modern computers. A procedure has been developed recently by Fraley and Rao[26] for the data obtained using double-pen recorders. This technique is described below.

Experimental Setup:

In order to clarify the discussion appearing later, a few salient details of the experimental setup are given in the next paragraph.

Two beams are passed through the entrance slit of a grating monochromator. One of these comprises the infrared under study and the other is from an emission source producing atomic lines. After dispersion, the first beam is focused on a sensitive infrared detector and the second on a photomultiplier or other appropriate detector. The signals from these detectors are amplified and recorded on the two pens of a double-pen recorder. The two traces so obtained may be referred to as the "infrared spectrum" and the "wave number markers." In view of the physical separation between the two pens of a double-pen recorder, the differences in the optical paths of the two beams, and other factors, the wave

[26] See Fraley and Rao.[24]

number scales of the infrared spectrum and the wave number markers are different. For a specific location on the chart, if ν_2 is the wave number determined from the wave number markers, the unknown wave number ν_1 of the infrared spectrum at the same position on the chart can be calculated if the multiplication factor $[F = (\nu_1/\nu_2)]$ is known.

Determination of F is complicated by the fact that it varies with the angular position of the grating. With the availability of infrared standards especially suitable for use with coarse gratings, the problem of determining F has become somewhat simpler. In the present discussion, absorption spectra will be considered in an illustration of this procedure.

Basic Observational Data:

A typical set of observational data recorded on the double-pen recorder consists of two charts.

Chart A: On this chart, infrared absorption spectra for which spectral positions are to be determined are traced by one pen of the double-pen recorder, and atomic emission lines are recorded by the second pen at the same time.

Chart B: This chart covers the same angular range of grating rotation used in obtaining chart A. Chosen absorption standards are traced by the pen employed for recoring the unknown infrared lines in chart A. At the same time, the atomic emission lines appearing in chart A are traced by the other pen. The selection of absorption standards is such that observations can be made at as many places as possible in the entire angular range covered in chart A. The absorbing gas giving rise to the infrared standards can be introduced into the spectrograph so that the optical path of the "infrared beam" is maintained while charts A and B are recorded. Also, the standards should be so chosen that the same infrared detector is used for obtaining the data in both charts.

Computational Procedure:

The distance of each of the lines recorded by both pens on charts A and B is measured from an arbitrary reference point P on each record. (P need not be the same for both charts.)

Chart B, on which infrared absorption standards are recorded, is considered first. The rotational lines of CO bands are used as absorption standards, and the measured distance (X) for each of the rotational lines of CO with the known wave number ν_{CO} is inserted in the following equation:

$$\nu_{CO} = A_1 + B_1 X + C_1 X^2 + D_1 X^3 \qquad (V\text{-}6)$$

A cubic equation of this type was shown to be useful after several preliminary trials. It should be emphasized that A_1, B_1, C_1, and D_1 are merely constants appearing in the cubic equation (V-6) and should not be confused with molecular parameters. When a coarse grating (for example, one with 73.25 grooves/mm) is used, the CO is usually observed in a higher spectral order. The ν_{CO} used here is the corresponding wave number in the first order of the grating. A least squares solution by computer gives values for A_1, B_1, C_1, and D_1; the calculated wave numbers of the rotational lines of CO bands obtained from these

constants should agree with the actual positions within preset limits of error. For data obtained with a high-resolution Ebert-type spectrograph[27] employing a precision screw to
guide the rotation of the grating, the rms deviation at 2.7 μ is within ± 0.007 cm^{-1} in most
cases. Whenever the deviation is much larger than this, it is usually the result of a measurement error or the blended nature of a particular CO line. A similar procedure can be
devised using wavelengths instead of wave numbers for spectrographs having grating drives
nearly linear in wavelength.

The next step is representation of the measurements for the atomic lines (neon, in
this example) on chart B by the relationship

$$\nu_{Ne} \text{ (chart B)} = A_2 + B_2 X + C_2 X^2 + D_2 X^3 \tag{V-7}$$

ν_{Ne} is again the first order wave number of the neon line actually occurring in a higher
spectral order.

Equation (V-6) can be used to calculate the wave number ν_{Ne} for each X corresponding to a measurement of a neon line. The multiplication factor, F, can then be determined
from the relationship

$$F = \frac{\nu_{Ne} \text{ (referred to CO standards)}}{\nu_{Ne}} \tag{V-8}$$

A value for F is calculated at each of the positions on chart B where a neon line has been
observed. This value of F is represented in the following equation

$$F = C_0 + C_1(\nu_{Ne}) + C_2(\nu_{Ne})^2 + C_3(\nu_{Ne})^3 \tag{V-9}$$

It is to be noted that F is determined as a function of ν_{Ne}.

The data from chart A are then considered. Measurements made for the neon lines
on this chart are inserted in an equation similar to (V-7)

$$\nu_{Ne} \text{ (chart A)} = A_2' + B_2' X' + C_2' X'^2 + D_2' X'^3 \tag{V-10}$$

In order to calculate the wave numbers of the unknown infrared lines in chart A, the measurement, X, for each line is used in Eq. (V-10) to calculate $\nu_{infrared}$ (as referred to
neon). Substituting this value for ν_{Ne} in Eq. (V-9) gives the appropriate multiplication
factor F_a. The correct wave number, ν, of the unknown infrared line is given by the
equation

$$\nu = [\nu_{ir} \text{ (as referred to neon)}] F_a \tag{V-11}$$

All operations indicated by Eqs. (V-6)-(V-11) can be performed in a single computer
program with different input parameters.

The expression of F as a function of ν_{Ne} in Eq. (V-9) has the unique advantage that
even if charts A and B are traced at slightly different recorder or grating speeds, the data
obtained are still usable if the drives are smooth. It may be necessary, however, to
"weight" the data according to the measuring accuracy expected for specific speeds of the
grating and recorder.

[27] See Rao and Nielsen.[8]

Additional Observational Data on Chart A:

At various predetermined spots on chart A, it is possible, by suitable manipulation of the foreprism monochromator, to record a few internal standards without stopping the rotation of the grating.[28] These internal standards can be used to calculate the spectral positions of some of the unknown infrared lines. These data give additional verifications of the validity of the procedure and can be handled by the same computer program by setting $F = 1.0$ (i. e. $C_1 = C_2 = C_3 = 0$ and $C_0 = 1$).

Fraley and Rao[29] chose the $2.7\ \mu$ bands of the H_2O molecule for a demonstration of the usefulness of the above technique. The water vapor spectrum extends over a wide wave number range ($250\text{-}300\ cm^{-1}$). The observations were accumulated during a period of several weeks, and the results obtained by employing the above calibration procedure were found to be satisfactory. In this method, the multiplication factor F includes the effect of the physical separation between the two pens of the double-pen recorder. It has been mentioned in an earlier section that Edwards et al. (p. 161) indicated that the dynamic pen separation (that is, the separation between the two pens of a double-pen recorder while it is recording data) is not exactly the same as the stationary pen separation. If records similar to those shown in Fig. 9b are obtained, it should be possible to determine the dynamic pen separation and remove its effect on F. The satisfactory nature of the results[29] obtained with the effect of the pen separation included in F indicates that even when the stationary pen separation is used to remove its influence on the factor F, the technique will still lead to reliable values for spectral positions in the infrared.

In this section, the usefulness of computer techniques has been illustrated with the use of visible neon lines for the production of the wave number scale on the infrared spectra to be measured. It has been mentioned earlier that it is sometimes necessary to use absorption lines as wave number markers, especially when one uses a coarse echelle, for example, one with 30 grooves/mm. In this case, the ν_{Ne} has to be replaced by the wave numbers of the absorption lines used for the wave number scale.

Concluding Comments:

A description of all the techniques currently employed in infrared calibrations has been presented with a view to providing a basis to which modifications and refinements suitable for a particular problem can be introduced. It is not implied that one or another of these techniques must be adopted in the exact manner discussed above for the determination of accurate spectral positions in the infrared.

[28] See Rao et al.[13]
[29] See Fraley and Rao.[24]

APPENDIX I

CONVERSION OF WAVELENGTHS IN AIR TO WAVE NUMBERS IN VACUUM AND VICE VERSA

The equation $\nu_{vac} = 1/n\lambda_{air}$ permits the conversion of a wavelength in standard air to a wave number in vacuum, provided the value of n, the refractive index for standard air, is known. (Standard air is dry air containing 0.03% by volume of CO_2 at normal pressure 760 mm Hg and having an air temperature of 15°C.) Edlén's[1] dispersion formula for standard air is given by

$$n_{std\ air} = 1 + 6432.8 \times 10^{-8} + \frac{2\,949\,810}{146 \times 10^8 - \nu^2} + \frac{25\,540}{41 \times 10^8 - \nu^2} \tag{I-A}$$

where ν is the wave number expressed in cm^{-1}. Based on the work of Edlén, extensive tables of wave numbers were prepared by Coleman, Bozman, and Meggers[2] for the wavelength from 2000 Å - 1000 μ. Prior to this publication, the work of Penndorf[3] dealt with the values for the refractive indexes of standard air for the spectral region between 0.2 and 20 μ.

Recently, Edlén prepared conversion tables for $\lambda > 10,000$ Å suitable for use with desk calculators. They are reproduced as Tables I.1 and I.2 of this Appendix, and a few examples are given to illustrate their use.

Table I.1: This table can be used for the conversion of air wavelengths to vacuum wave numbers (expressed in cm^{-1}). For example (Table I, Chapter I) in order to convert $\lambda_{air} = 12802.737$ Å (12802.737×10^{-8} cm) to ν_{vac} (cm^{-1}), we note that 12802.737 (λ_{air}) occurs between the numbers 12740 and 12807 in Table I.1. The vertical line between these two numbers terminates in the digit 3. At the left, in column 1, we find the reciprocal index of standard air (n^{-1}) to be 0.9997265. To this value we link the digit 3 so that it becomes 0.99972653; therefore

$$\nu_{vac} = \frac{0.99972653}{12802.737 \times 10^{-8}} = 7808.694 \ cm^{-1}$$

Table I.2: This table is to be used to obtain λ_{air} from ν_{vac}. For example, if $\nu_{vac} = 7287.393$ cm^{-1}, this value can be located between the numbers 7293 and 7249 in Table I.2.

[1] B. Edlén, *J. Opt. Soc. Am.* **43**, 339 (1953).
[2] C. D. Coleman, W. R. Bozman, and W. F. Meggers, *Natl. Bur. Std. (U.S.), Monograph* **3**, Vols. I and II (1960).
[3] R. Penndorf, *J. Opt. Soc. Am.* **47**, 176 (1957).

The vertical line between these two numbers terminates in 6 and the horizontal line connecting them leads to 0.9997266 for n^{-1}. Proceeding as above,

$$\lambda_{air} = \frac{0.99972666}{7287.393 \times 10^{-8}} = 13718.577 \text{ Å}$$

In other words, the numbers given in both tables help us only to locate the last digit for n^{-1}. No interpolations are necessary. In Table I.1 the numbers following $\lambda 20108$ Å are given in microns.

It should be recalled that it is important to determine the wavelength correction required for nonstandard air conditions. The following formula from Edlén can be used:

$$\lambda_2^0 - \lambda_2 = \left[\Delta\lambda_2 - \Delta\lambda_1 (\lambda_2/\lambda_1) \right] \left[\frac{0.0013882 p}{1 + 0.00367 t} - 1 \right] \tag{I-B}$$

where λ_2^0 = unknown wavelength at standard temperature and pressure

λ_2 = unknown wavelength as measured in nonstandard air

λ_1 = reference wavelength (for instance, if the green radiation from Hg^{198} is used, $\lambda_1 = 5460.7529$ Å in standard air)

and $\dfrac{\Delta\lambda_2}{\Delta\lambda_1}$ = vacuum corrections for λ_2 and λ_1 (for instance, the vacuum correction for Hg^{198} green line is 1.5175 Å)

p = atmospheric pressure in millimeters of Hg

t = temperature in °C

In high precision work, an additional correction should be made for the water vapor. This is given by

$$\lambda_2^0 - \lambda_2 = [+0.63 \ (1 + \lambda_2/\lambda_1) \ (\nu_2 - \nu_1)f] \times 10^{-9} \tag{I-C}$$

where $\lambda_2^0 - \lambda_2$ and λ_1 are expressed in angstroms and ν_2 and ν_1 in cm^{-1}; f is in millimeters of Hg of water vapor. For instance, if $\lambda_2 = 4046$ Å, $\lambda_1 = 5460$ Å, and $f = 10$ mm of Hg, the correction due to water vapor is $+0.00007$ Å.

TABLE I. 1

Reciprocal Refractive Index of Standard Air for $\lambda > 10{,}000$ Å ^{a}

n^{-1}		0	1	2	3	4	5	6	7	8	9
0.99972 59	λ_{air}:	9928	9959	9991	10023	10056	10088	10121	10155	10188	10222
60	10222	10257	10292	10327	10362	10398	10434	10471	10508	10546	10583
61	10583	10622	10660	10699	10739	10779	10819	10860	10902	10944	10986
62	10986	11029	11072	11116	11160	11205	11251	11297	11344	11391	11439
63	11439	11487	11536	11586	11636	11687	11739	11791	11845	11898	11953
64	11953	12008	12065	12122	12179	12238	12297	12358	12419	12481	12544
65	12544	12609	12674	12740	12807	12875	12945	13015	13087	13160	13234
66	13234	13310	13386	13464	13544	13625	13707	13791	13877	13964	14053
67	14053	14143	14236	14330	14426	14524	14624	14726	14830	14937	15046
68	15046	15157	15271	15388	15507	15629	15754	15882	16014	16148	16286
69	16286	16428	16574	16723	16877	17035	17197	17364	17537	17714	17897
70	17897	18086	18281	18482	18690	18906	19129	19360	19600	19849	20108
71	2.011	2.038	2.066	2.095	2.125	2.157	2.191	2.226	2.262	2.301	2.342
72	2.342	2.384	2.430	2.478	2.529	2.583	2.641	2.703	2.769	2.841	2.919
73	2.919	3.003	3.096	3.197	3.309	3.433	3.573	3.732	3.913	4.124	4.373
74	4.373	4.674	5.047	5.527	6.175	7.123	8.706	12.24	79.8	∞	
		0	1	2	3	4	5	6	7	8	9

^{a} $\lambda_{air} = n^{-1}/\nu_{vac}$; $\nu_{vac} = n^{-1}/\lambda_{air}$.

TABLE I. 2

Reciprocal Refractive Index of Standard Air for $\nu < 10{,}000$ cm^{-1} [a]

n^{-1}	ν_{vac}:	0	1	2	3	4	5	6	7	8	9
0.99972 59		10070	10038	10006	9974	9942	9910	9877	9845	9812	9780
60	9780	9747	9714	9681	9648	9614	9581	9547	9514	9480	9446
61	9446	9412	9378	9344	9309	9275	9240	9205	9170	9135	9100
62	9100	9065	9029	8994	8958	8922	8886	8850	8813	8777	8740
63	8740	8703	8666	8629	8591	8554	8516	8478	8440	8402	8364
64	8364	8325	8286	8247	8208	8169	8130	8090	8050	8010	7969
65	7969	7929	7888	7847	7806	7765	7723	7681	7639	7597	7554
66	7554	7511	7468	7425	7381	7337	7293	7249	7204	7159	7114
67	7114	7069	7023	6977	6930	6883	6836	6789	6741	6693	6644
68	6644	6596	6546	6497	6447	6396	6346	6295	6243	6191	6138
69	6138	6085	6032	5978	5924	5869	5813	5757	5701	5644	5586
70	5586	5528	5469	5409	5349	5288	5226	5164	5101	5037	4972
71	4972	4906	4840	4772	4704	4634	4564	4492	4419	4345	4270
72	4270	4193	4115	4035	3954	3870	3786	3699	3610	3519	3425
73	3425	3329	3230	3127	3021	2912	2798	2679	2555	2424	2286
74	2286	2139	1981	1809	1619	1404	1148	817	125	0	
		0	1	2	3	4	5	6	7	8	9

[a] $\lambda_{\text{air}} = n^{-1}/\nu_{\text{vac}}$; $\nu_{\text{vac}} = n^{-1}/\lambda_{\text{air}}$.

APPENDIX II

TABLE II. 1

Molecular Constants Used in Obtaining "Calculated" Values of the Rotational Lines of the Bands Listed in Chapter III

Molecule and band / Rotational constants (cm^{-1})	$HC^{12}N^{14}$ ($2\nu_3$)	$N_2^{14}O^{16}$ ($3\nu_1 + \nu_3$)	$HC^{12}N^{14}$ ($\nu_1 + \nu_3$)	$N_2^{14}O^{16}$ ($2\nu_1 + \nu_3$)	$N_2^{14}O^{16}$ ($\nu_1 + 2\nu_2 + \nu_3$)	$HC^{12}N^{14}$ (ν_3)	$N_2^{14}O^{16}$ ($2\nu_1$)	$N_2^{14}O^{16}$ ($\nu_1 + 2\nu_2$)[a]
B''	1.478219	0.4190107	1.478219	0.4190107	0.4190107	1.478219	0.4190107	0.4190107
B'	1.457070	0.4106660	1.457933	0.4121167	0.4147520	1.467798	0.4156031	0.4181457
D''[b]	2.9137×10^{-6}	1.757×10^{-7}	2.9137×10^{-6}	1.757×10^{-7}	1.757×10^{-7}	2.9129×10^{-6}	1.757×10^{-7}	1.757×10^{-7}
D'[b]	2.8637×10^{-6}	1.332×10^{-7}	2.8995×10^{-6}	1.594×10^{-7}	2.310×10^{-7}	2.8890×10^{-6}	1.608×10^{-7}	2.381×10^{-7}
$H_v \approx H_e$	2.7×10^{-12}	8×10^{-14}	2.7×10^{-12}	8×10^{-14}	8×10^{-14}	5.73×10^{-12}	8×10^{-14}	8×10^{-14}
ν_0	6519.6135	5974.8504	5393.698	4730.8274	4630.1659	3311.4776	2563.3408	2461.9981

[a] The molecular constants for this band were determined from combination differences up to $J = 35$. The "calculated" values of the wave numbers for the rotational lines of this band given on pp. 126–131 were obtained by taking account of the perturbation of the upper state of this band.

[b] It is to be noted that in some of the publications of Rank *et al.* the distortion constants appear with negative signs because these authors found it convenient to express rotational term values by $F(J) = BJ(J+1) + DJ^2(J+1)^2 + HJ^3(J+1)^3 \cdots$, instead of $F(J) = BJ(J+1) - DJ^2(J+1)^2 + HJ^3(J+1)^3 \cdots$. The notation appearing in the above table follows the recommendations of R. S. Mulliken, *J. Chem. Phys.* 23, 1997 (1955).

181

TABLE II.2

Molecular Constants Used in Obtaining "Calculated" Values of the Rotational Lines
of the CO Bands (cm⁻¹) Listed in Chapter III

Band origins
$\nu_0 (1 \to 0) = 2143.2731$
$\nu_0 (2 \to 0) = 4260.0646$

Rotational constants		
$B_2 = 1.887513$	$D_2 = 6.1174 \times 10^{-6}$	$H_2 \approx H_1 \approx H_0 = 5.8 \times 10^{-12}$
$B_1 = 1.905014$	$D_1 = 6.1183 \times 10^{-6}$	
$B_0 = 1.922521$	$D_0 = 6.1193 \times 10^{-6}$ [a]	

[a] Recently, G. Jones and W. Gordy [*Phys. Rev.* **135**, 295 (1964)], determined a $D_0 = 6.1342 \times 10^{-6}$ cm⁻¹ which is slightly different from the D_0 quoted above. This difference does not affect the wave numbers quoted in this monograph for the CO lines between R (30) and P (30). To be sure, the CO standards given in this monograph are satisfactory to at least ± 0.002 cm⁻¹.

FIRST ORDER WAVE NUMBERS OF MOLECULAR ABSORPTION
STANDARDS FOR COARSE ECHELLES

APPENDIX III. 1

First Order Wave Numbers of the Rotational Lines of the 1-0 Band of $C^{12}O^{16}$

1-0 band of CO occurring in the fourth order

Line J	First Order Wave Number (vac cm^{-1})	$\Delta \nu$	Line J	First Order Wave Number (vac cm^{-1})	$\Delta \nu$
R 30	561.09393		R 11	546.66008	
		0.67290			0.85380
R 29	560.42103		R 10	545.80628	
		0.68275			0.86295
R 28	559.73828		R 9	544.94333	
		0.69255			0.87208
R 27	559.04573		R 8	544.07125	
		0.70235			0.88115
R 26	558.34338		R 7	543.19010	
		0.71208			0.89020
R 25	557.63130		R 6	542.29990	
		0.72177			0.89920
R 24	556.90953		R 5	541.40070	
		0.73143			0.90820
R 23	556.17810		R 4	540.49250	
		0.74107			0.91710
R 22	555.43703		R 3	539.57540	
		0.75068			0.92602
R 21	554.68635		R 2	538.64938	
		0.76022			0.93490
R 20	553.92613		R 1	537.71448	
		0.76973			0.94370
R 19	553.15640		R 0	536.77078	
		0.77922			1.91375
R 18	552.37718		P 1	534.85703	
		0.78868			0.96998
R 17	551.58850		P 2	533.88705	
		0.79810			0.97865
R 16	550.79040		P 3	532.90840	
		0.80747			0.98730
R 15	549.98293		P 4	531.92110	
		0.81680			0.99590
R 14	549.16613		P 5	530.92520	
		0.82610			1.00447
R 13	548.34003		P 6	529.92073	
		0.83535			1.01300
R 12	547.50468		P 7	528.90773	
		0.84460			1.02150
R 11	546.66008		P 8	527.88623	

1-0 band of CO occurring in the fourth order

Line J	First Order Wave Number (vac cm⁻¹)	Δy	Line J	First Order Wave Number (vac cm⁻¹)	Δy
P 8	527.88623		P 19	516.09948	
		1.02995			1.12060
P 9	526.85628		P 20	514.97888	
		1.03840			1.12860
P 10	525.81788		P 21	513.85028	
		1.04675			1.13660
P 11	524.77113		P 22	512.71368	
		1.05513			1.14453
P 12	523.71600		P 23	511.56915	
		1.06342			1.15245
P 13	522.65258		P 24	510.41670	
		1.07173			1.16032
P 14	521.58085		P 25	509.25638	
		1.07992			1.16815
P 15	520.50093		P 26	508.08823	
		1.08815			1.17595
P 16	519.41278		P 27	506.91228	
		1.09633			1.18373
P 17	518.31645		P 28	505.72855	
		1.10445			1.19145
P 18	517.21200		P 29	504.53710	
		1.11252			1.19912
P 19	516.09948		P 30	503.33798	

First Order Wave Numbers of the Rotational Lines of the 1-0 Band of C¹²O¹⁶

1-0 band of CO occurring in the fifth order

Line J	First Order Wave Number (vac cm^{-1})	Δν	Line J	First Order Wave Number (vac cm^{-1})	Δν
R 30	448.87514		R 11	437.32806	
		0.53832			0.68304
R 29	448.33682		R 10	436.64502	
		0.54620			0.69036
R 28	447.79062		R 9	435.95466	
		0.55404			0.69766
R 27	447.23658		R 8	435.25700	
		0.56188			0.70492
R 26	446.67470		R 7	434.55208	
		0.56966			0.71216
R 25	446.10504		R 6	433.83992	
		0.57742			0.71936
R 24	445.52762		R 5	433.12056	
		0.58514			0.72656
R 23	444.94248		R 4	432.39400	
		0.59286			0.73368
R 22	444.34962		R 3	431.66032	
		0.60054			0.74082
R 21	443.74908		R 2	430.91950	
		0.60818			0.74792
R 20	443.14090		R 1	430.17158	
		0.61578			0.75496
R 19	442.52512		R 0	429.41662	
		0.62338			1.53100
R 18	441.90174		P 1	427.88562	
		0.63094			0.77598
R 17	441.27080		P 2	427.10964	
		0.63848			0.78292
R 16	440.63232		P 3	426.32672	
		0.64598			0.78984
R 15	439.98634		P 4	425.53688	
		0.65344			0.79672
R 14	439.33290		P 5	424.74016	
		0.66088			0.80358
R 13	438.67202		P 6	423.93658	
		0.66828			0.81040
R 12	438.00374		P 7	423.12618	
		0.67568			0.81720
R 11	437.32806		P 8	422.30898	

185

1-0 band of CO occurring in the fifth order

Line J	First Order Wave Number (vac cm^{-1})	Δy	Line J	First Order Wave Number (vac cm^{-1})	Δy
P 8	422.30898		P 19	412.87958	
		0.82396			0.89648
P 9	421.48502		P 20	411.98310	
		0.83072			0.90288
P 10	420.65430		P 21	411.08022	
		0.83740			0.90928
P 11	419.81690		P 22	410.17094	
		0.84410			0.91562
P 12	418.97280		P 23	409.25532	
		0.85074			0.92196
P 13	418.12206		P 24	408.33336	
		0.85738			0.92826
P 14	417.26468		P 25	407.40510	
		0.86394			0.93452
P 15	416.40074		P 26	406.47058	
		0.87052			0.94076
P 16	415.53022		P 27	405.52982	
		0.87706			0.94698
P 17	414.65316		P 28	404.58284	
		0.88356			0.95316
P 18	413.76960		P 29	403.62968	
		0.89002			0.95930
P 19	412.87958		P 30	402.67038	

First Order Wave Numbers of the Rotational Lines of the 1-0 Band of $C^{13}O^{16}$

1-0 band of CO occurring in the sixth order

Line J	First Order Wave Number (vac cm^{-1})	$\Delta\nu$	Line J	First Order Wave Number (vac cm^{-1})	$\Delta\nu$
R 30	374.06262		R 11	364.44005	
		0.44860			0.56920
R 29	373.61402		R 10	363.87085	
		0.45517			0.57530
R 28	373.15885		R 9	363.29555	
		0.46170			0.58138
R 27	372.69715		R 8	362.71417	
		0.46823			0.58744
R 26	372.22892		R 7	362.12673	
		0.47472			0.59346
R 25	371.75420		R 6	361.53327	
		0.48118			0.59947
R 24	371.27302		R 5	360.93380	
		0.48762			0.60547
R 23	370.78540		R 4	360.32833	
		0.49405			0.61140
R 22	370.29135		R 3	359.71693	
		0.50045			0.61735
R 21	369.79090		R 2	359.09958	
		0.50682			0.62326
R 20	369.28408		R 1	358.47632	
		0.51315			0.62914
R 19	368.77093		R 0	357.84718	
		0.51948			1.27583
R 18	368.25145		P 1	356.57135	
		0.52578			0.64665
R 17	367.72567		P 2	355.92470	
		0.53207			0.65243
R 16	367.19360		P 3	355.27227	
		0.53832			0.65820
R 15	366.65528		P 4	354.61407	
		0.54453			0.66394
R 14	366.11075		P 5	353.95013	
		0.55073			0.66965
R 13	365.56002		P 6	353.28048	
		0.55690			0.67533
R 12	365.00312		P 7	352.60515	
		0.56307			0.68100
R 11	364.44005		P 8	351.92415	

1-0 band of CO occurring in the sixth order

Line J	First Order Wave Number (vac cm^{-1})	$\Delta\nu$	Line J	First Order Wave Number (vac cm^{-1})	$\Delta\nu$
P 8	351.92415		P 19	344.06632	
		0.68663			0.74707
P 9	351.23752		P 20	343.31925	
		0.69227			0.75240
P 10	350.54525		P 21	342.56685	
		0.69783			0.75773
P 11	349.84742		P 22	341.80912	
		0.70342			0.76302
P 12	349.14400		P 23	341.04610	
		0.70895			0.76830
P 13	348.43505		P 24	340.27780	
		0.71448			0.77355
P 14	347.72057		P 25	339.50425	
		0.71995			0.77877
P 15	347.00062		P 26	338.72548	
		0.72544			0.78396
P 16	346.27518		P 27	337.94152	
		0.73088			0.78915
P 17	345.54430		P 28	337.15237	
		0.73630			0.79430
P 18	344.80800		P 29	336.35807	
		0.74168			0.79942
P 19	344.06632		P 30	335.55865	

First Order Wave Numbers of the Rotational Lines of the 1-0 Band of $C^{12}O^{16}$

1-0 band of CO occurring in the seventh order

Line J	First Order Wave Number (vac cm^{-1})	$\Delta\nu$	Line J	First Order Wave Number (vac cm^{-1})	$\Delta\nu$
R 30	320.62510		R 11	312.37719	
		0.38451			0.48789
R 29	320.24059		R 10	311.88930	
		0.39015			0.49311
R 28	319.85044		R 9	311.39619	
		0.39574			0.49833
R 27	319.45470		R 8	310.89786	
		0.40134			0.50352
R 26	319.05336		R 7	310.39434	
		0.40690			0.50868
R 25	318.64646		R 6	309.88566	
		0.41245			0.51383
R 24	318.23401		R 5	309.37183	
		0.41795			0.51897
R 23	317.81606		R 4	308.85286	
		0.42347			0.52406
R 22	317.39259		R 3	308.32880	
		0.42896			0.52916
R 21	316.96363		R 2	307.79964	
		0.43442			0.53423
R 20	316.52921		R 1	307.26541	
		0.43984			0.53925
R 19	316.08937		R 0	306.72616	
		0.44527			1.09357
R 18	315.64410		P 1	305.63259	
		0.45067			0.55428
R 17	315.19343		P 2	305.07831	
		0.45606			0.55922
R 16	314.73737		P 3	304.51909	
		0.46141			0.56418
R 15	314.27596		P 4	303.95491	
		0.46675			0.56908
R 14	313.80921		P 5	303.38583	
		0.47205			0.57399
R 13	313.33716		P 6	302.81184	
		0.47735			0.57885
R 12	312.85981		P 7	302.23299	
		0.48262			0.58372
R 11	312.37719		P 8	301.64927	

1-0 band of CO occurring in the seventh order

Line J	First Order Wave Number (vac cm-1)	$\Delta\nu$	Line J	First Order Wave Number (vac cm-1)	$\Delta\nu$
P 8	301.64927		P 19	294.91399	
		0.58854			0.64035
P 9	301.06073		P 20	294.27364	
		0.59337			0.64491
P 10	300.46736		P 21	293.62873	
		0.59815			0.64949
P 11	299.86921		P 22	292.97924	
		0.60292			0.65401
P 12	299.26629		P 23	292.32523	
		0.60768			0.65854
P 13	298.65861		P 24	291.66669	
		0.61241			0.66305
P 14	298.04620		P 25	291.00364	
		0.61710			0.66751
P 15	297.42910		P 26	290.33613	
		0.62180			0.67197
P 16	296.80730		P 27	289.66416	
		0.62647			0.67642
P 17	296.18083		P 28	288.98774	
		0.63112			0.68083
P 18	295.54971		P 29	288.30691	
		0.63572			0.68521
P 19	294.91399		P 30	287.62170	

First Order Wave Numbers of the Rotational Lines of the 1-0 Band of $C^{12}O^{16}$

1-0 band of CO occurring in the eighth order

Line J	First Order Wave Number (vac cm^{-1})	$\Delta\nu$	Line J	First Order Wave Number (vac cm^{-1})	$\Delta\nu$
R 30	280.54696		R 11	273.33004	
		0.33645			0.42690
R 29	280.21051		R 10	272.90314	
		0.34137			0.43148
R 28	279.86914		R 9	272.47166	
		0.34628			0.43603
R 27	279.52286		R 8	272.03563	
		0.35117			0.44058
R 26	279.17169		R 7	271.59505	
		0.35604			0.44510
R 25	278.81565		R 6	271.14995	
		0.36089			0.44960
R 24	278.45476		R 5	270.70035	
		0.36571			0.45410
R 23	278.08905		R 4	270.24625	
		0.37054			0.45855
R 22	277.71851		R 3	269.78770	
		0.37533			0.46301
R 21	277.34318		R 2	269.32469	
		0.38012			0.46745
R 20	276.96306		R 1	268.85724	
		0.38486			0.47185
R 19	276.57820		R 0	268.38539	
		0.38961			0.95688
R 18	276.18859		P 1	267.42851	
		0.39434			0.48498
R 17	275.79425		P 2	266.94353	
		0.39905			0.48933
R 16	275.39520		P 3	266.45420	
		0.40374			0.49365
R 15	274.99146		P 4	265.96055	
		0.40840			0.49795
R 14	274.58306		P 5	265.46260	
		0.41305			0.50224
R 13	274.17001		P 6	264.96036	
		0.41767			0.50650
R 12	273.75234		P 7	264.45386	
		0.42230			0.51075
R 11	273.33004		P 8	263.94311	

1-0 band of CO occurring in the eighth order

Line J	First Order Wave Number (vac cm⁻¹)	$\Delta \nu$	Line J	First Order Wave Number (vac cm⁻¹)	$\Delta \nu$
P 8	263.94311		P 19	258.04974	
		0.51497			0.56030
P 9	263.42814		P 20	257.48944	
		0.51920			0.56430
P 10	262.90894		P 21	256.92514	
		0.52338			0.56830
P 11	262.38556		P 22	256.35684	
		0.52756			0.57226
P 12	261.85800		P 23	255.78458	
		0.53171			0.57623
P 13	261.32629		P 24	255.20835	
		0.53586			0.58016
P 14	260.79043		P 25	254.62819	
		0.53997			0.58408
P 15	260.25046		P 26	254.04411	
		0.54407			0.58797
P 16	259.70639		P 27	253.45614	
		0.54816			0.59186
P 17	259.15823		P 28	252.86428	
		0.55223			0.59573
P 18	258.60600		P 29	252.26855	
		0.55626			0.59956
P 19	258.04974		P 30	251.66899	

APPENDIX III. 6

First Order Wave Numbers of the Rotational Lines of the 1-0 Band of $C^{12}O^{16}$

1-0 band of CO occurring in the ninth order

Line J	First Order Wave Number (vac cm^{-1})	$\Delta\nu$	Line J	First Order Wave Number (vac cm^{-1})	$\Delta\nu$
R 30	249.37508		R 11	242.96003	
		0.29907			0.37946
R 29	249.07601		R 10	242.58057	
		0.30344			0.38354
R 28	248.77257		R 9	242.19703	
		0.30780			0.38759
R 27	248.46477		R 8	241.80944	
		0.31216			0.39162
R 26	248.15261		R 7	241.41782	
		0.31648			0.39564
R 25	247.83613		R 6	241.02218	
		0.32079			0.39965
R 24	247.51534		R 5	240.62253	
		0.32507			0.40364
R 23	247.19027		R 4	240.21889	
		0.32937			0.40760
R 22	246.86090		R 3	239.81129	
		0.33363			0.41157
R 21	246.52727		R 2	239.39972	
		0.33788			0.41551
R 20	246.18939		R 1	238.98421	
		0.34210			0.41942
R 19	245.84729		R 0	238.56479	
		0.34632			0.85056
R 18	245.50097		P 1	237.71423	
		0.35053			0.43110
R 17	245.15044		P 2	237.28313	
		0.35471			0.43495
R 16	244.79573		P 3	236.84818	
		0.35887			0.43880
R 15	244.43686		P 4	236.40938	
		0.36303			0.44262
R 14	244.07383		P 5	235.96676	
		0.36715			0.44644
R 13	243.70668		P 6	235.52032	
		0.37127			0.45022
R 12	243.33541		P 7	235.07010	
		0.37538			0.45400
R 11	242.96003		P 8	234.61610	

1-0 band of CO occurring in the ninth order

Line J	First Order Wave Number (vac cm⁻¹)	Δν	Line J	First Order Wave Number (vac cm⁻¹)	Δν
P 8	234.61610		P 19	229.37754	
		0.45776			0.49804
P 9	234.15834		P 20	228.87950	
		0.46151			0.50160
P 10	233.69683		P 21	228.37790	
		0.46522			0.50516
P 11	233.23161		P 22	227.87274	
		0.46894			0.50867
P 12	232.76267		P 23	227.36407	
		0.47264			0.51220
P 13	232.29003		P 24	226.85187	
		0.47632			0.51570
P 14	231.81371		P 25	226.33617	
		0.47997			0.51918
P 15	231.33374		P 26	225.81699	
		0.48362			0.52265
P 16	230.85012		P 27	225.29434	
		0.48725			0.52610
P 17	230.36287		P 28	224.76824	
		0.49087			0.52953
P 18	229.87200		P 29	224.23871	
		0.49446			0.53294
P 19	229.37754		P 30	223.70577	

First Order Wave Numbers of the Rotational Lines of the 1-0 Band of $C^{12}O^{16}$

1-0 band of CO occurring in the eleventh order

Line J	First Order Wave Number (vac cm^{-1})	$\Delta\nu$	Line J	First Order Wave Number (vac cm^{-1})	$\Delta\nu$
R 30	204.03415		R 11	198.78548	
		0.24469			0.31047
R 29	203.78946		R 10	198.47501	
		0.24827			0.31380
R 28	203.54119		R 9	198.16121	
		0.25184			0.31712
R 27	203.28935		R 8	197.84409	
		0.25540			0.32042
R 26	203.03395		R 7	197.52367	
		0.25893			0.32371
R 25	202.77502		R 6	197.19996	
		0.26247			0.32698
R 24	202.51255		R 5	196.87298	
		0.26597			0.33025
R 23	202.24658		R 4	196.54273	
		0.26948			0.33349
R 22	201.97710		R 3	196.20924	
		0.27297			0.33674
R 21	201.70413		R 2	195.87250	
		0.27645			0.33996
R 20	201.42768		R 1	195.53254	
		0.27990			0.34317
R 19	201.14778		R 0	195.18937	
		0.28335			0.69591
R 18	200.86443		P 1	194.49346	
		0.28679			0.35271
R 17	200.57764		P 2	194.14075	
		0.29022			0.35588
R 16	200.28742		P 3	193.78487	
		0.29363			0.35902
R 15	199.99379		P 4	193.42585	
		0.29702			0.36214
R 14	199.69677		P 5	193.06371	
		0.30040			0.36526
R 13	199.39637		P 6	192.69845	
		0.30376			0.36837
R 12	199.09261		P 7	192.33008	
		0.30713			0.37145
R 11	198.78548		P 8	191.95863	

1-0 band of CO occurring in the eleventh order

Line J	First Order Wave Number (vac cm⁻¹)	$\Delta \nu$	Line J	First Order Wave Number (vac cm⁻¹)	$\Delta \nu$
P 8	191.95863		P 19	187.67254	
		0.37453			0.40749
P 9	191.58410		P 20	187.26505	
		0.37760			0.41040
P 10	191.20650		P 21	186.85465	
		0.38064			0.41331
P 11	190.82586		P 22	186.44134	
		0.38368			0.41619
P 12	190.44218		P 23	186.02515	
		0.38670			0.41908
P 13	190.05548		P 24	185.60607	
		0.38972			0.42194
P 14	189.66576		P 25	185.18413	
		0.39270			0.42478
P 15	189.27306		P 26	184.75935	
		0.39569			0.42761
P 16	188.87737		P 27	184.33174	
		0.39866			0.43045
P 17	188.47871		P 28	183.90129	
		0.40162			0.43325
P 18	188.07709		P 29	183.46804	
		0.40455			0.43605
P 19	187.67254		P 30	183.03199	

First Order Wave Numbers of the Rotational Lines of the 1-0 Band of $C^{12}O^{16}$

1-0 band of CO occurring in the twelfth order

Line J	First Order Wave Number (vac cm^{-1})	$\Delta\nu$	Line J	First Order Wave Number (vac cm^{-1})	$\Delta\nu$
R 30	187.03131		R 11	182.22002	
		0.22430			0.28460
R 29	186.80701		R 10	181.93542	
		0.22758			0.28765
R 28	186.57943		R 9	181.64777	
		0.23085			0.29069
R 27	186.34858		R 8	181.35708	
		0.23412			0.29371
R 26	186.11446		R 7	181.06337	
		0.23736			0.29674
R 25	185.87710		R 6	180.76663	
		0.24059			0.29973
R 24	185.63651		R 5	180.46690	
		0.24381			0.30273
R 23	185.39270		R 4	180.16417	
		0.24703			0.30570
R 22	185.14567		R 3	179.85847	
		0.25022			0.30868
R 21	184.89545		R 2	179.54979	
		0.25341			0.31163
R 20	184.64204		R 1	179.23816	
		0.25657			0.31457
R 19	184.38547		R 0	178.92359	
		0.25975			0.63792
R 18	184.12572		P 1	178.28567	
		0.26289			0.32332
R 17	183.86283		P 2	177.96235	
		0.26603			0.32622
R 16	183.59680		P 3	177.63613	
		0.26916			0.32910
R 15	183.32764		P 4	177.30703	
		0.27227			0.33196
R 14	183.05537		P 5	176.97507	
		0.27536			0.33483
R 13	182.78001		P 6	176.64024	
		0.27845			0.33767
R 12	182.50156		P 7	176.30257	
		0.28154			0.34050
R 11	182.22002		P 8	175.96207	

1-0 band of CO occurring in the twelfth order

Line J	First Order Wave Number (vac cm^{-1})	$\Delta\nu$	Line J	First Order Wave Number (vac cm^{-1})	$\Delta\nu$
P 8	175.96207		P 19	172.03316	
		0.34331			0.37354
P 9	175.61876		P 20	171.65962	
		0.34614			0.37620
P 10	175.27262		P 21	171.28342	
		0.34891			0.37886
P 11	174.92371		P 22	170.90456	
		0.35171			0.38151
P 12	174.57200		P 23	170.52305	
		0.35448			0.38415
P 13	174.21752		P 24	170.13890	
		0.35724			0.38678
P 14	173.86028		P 25	169.75212	
		0.35997			0.38938
P 15	173.50031		P 26	169.36274	
		0.36272			0.39198
P 16	173.13759		P 27	168.97076	
		0.36544			0.39458
P 17	172.77215		P 28	168.57618	
		0.36815			0.39715
P 18	172.40400		P 29	168.17903	
		0.37084			0.39971
P 19	172.03316		P 30	167.77932	

First Order Wave Numbers of the Rotational Lines of the 1-0 Band of $C^{12}O^{16}$

1-0 band of CO occurring in the thirteenth order

Line J	First Order Wave Number (vac cm^{-1})	$\Delta\nu$	Line J	First Order Wave Number (vac cm^{-1})	$\Delta\nu$
R 30	172.64428		R 11	168.20310	
		0.20704			0.26271
R 29	172.43724		R 10	167.94039	
		0.21008			0.26552
R 28	172.22716		R 9	167.67487	
		0.21309			0.26833
R 27	172.01407		R 8	167.40654	
		0.21611			0.27112
R 26	171.79796		R 7	167.13542	
		0.21910			0.27391
R 25	171.57886		R 6	166.86151	
		0.22208			0.27668
R 24	171.35678		R 5	166.58483	
		0.22506			0.27945
R 23	171.13172		R 4	166.30538	
		0.22802			0.28218
R 22	170.90370		R 3	166.02320	
		0.23098			0.28493
R 21	170.67272		R 2	165.73827	
		0.23391			0.28766
R 20	170.43881		R 1	165.45061	
		0.23684			0.29037
R 19	170.20197		R 0	165.16024	
		0.23976			0.58885
R 18	169.96221		P 1	164.57139	
		0.24267			0.29845
R 17	169.71954		P 2	164.27294	
		0.24557			0.30112
R 16	169.47397		P 3	163.97182	
		0.24845			0.30379
R 15	169.22552		P 4	163.66803	
		0.25133			0.30643
R 14	168.97419		P 5	163.36160	
		0.25418			0.30907
R 13	168.72001		P 6	163.05253	
		0.25703			0.31169
R 12	168.46298		P 7	162.74084	
		0.25988			0.31431
R 11	168.20310		P 8	162.42653	

1-0 band of CO occurring in the thirteenth order

Line J	First Order Wave Number (vac cm⁻¹)	$\Delta\nu$	Line J	First Order Wave Number (vac cm⁻¹)	$\Delta\nu$
P 8	162.42653		P 19	158.79984	
		0.31691			0.34430
P 9	162.10962		P 20	158.45504	
		0.31950			0.34726
P 10	161.79012		P 21	158.10778	
		0.32208			0.34973
P 11	161.46804		P 22	157.75805	
		0.32466			0.35216
P 12	161.14338		P 23	157.40589	
		0.32720			0.35460
P 13	160.81618		P 24	157.05129	
		0.32976			0.35702
P 14	160.48642		P 25	156.69427	
		0.33229			0.35943
P 15	160.15413		P 26	156.33484	
		0.33481			0.36183
P 16	159.81932		P 27	155.97301	
		0.33734			0.36423
P 17	159.48198		P 28	155.60878	
		0.33983			0.36660
P 18	159.14215		P 29	155.24218	
		0.34231			0.36896
P 19	158.79984		P 30	154.87322	

First Order Wave Numbers of the Rotational Lines of the $2\nu_3$ Band of $HC^{12}N^{14}$

$2\nu_3$ band of HCN occurring in the fourteenth order

Line J	First Order Wave Number (vac cm^{-1})	$\Delta\nu$	Line J	First Order Wave Number (vac cm^{-1})	$\Delta\nu$
R 32	470.93511		R 13	468.32381	
		0.10935			0.16845
R 31	470.82576		R 12	468.15536	
		0.11248			0.17154
R 30	470.71328		R 11	467.98382	
		0.11562			0.17461
R 29	470.59766		R 10	467.80921	
		0.11875			0.17769
R 28	470.47891		R 9	467.63152	
		0.12187			0.18074
R 27	470.35704		R 8	467.45078	
		0.12500			0.18382
R 26	470.23204		R 7	467.26696	
		0.12813			0.18686
R 25	470.10391		R 6	467.08010	
		0.13125			0.18993
R 24	469.97266		R 5	466.89017	
		0.13436			0.19297
R 23	469.83830		R 4	466.69720	
		0.13748			0.19602
R 22	469.70082		R 3	466.50118	
		0.14060			0.19906
R 21	469.56022		R 2	466.30212	
		0.14370			0.20209
R 20	469.41652		R 1	466.10003	
		0.14681			0.20513
R 19	469.26971		R 0	465.89490	
		0.14990			0.41932
R 18	469.11981		P 1	465.47558	
		0.15302			0.21419
R 17	468.96679		P 2	465.26139	
		0.15610			0.21720
R 16	468.81069		P 3	465.04419	
		0.15921			0.22021
R 15	468.65148		P 4	464.82398	
		0.16229			0.22321
R 14	468.48919		P 5	464.60077	
		0.16538			0.22620
R 13	468.32381		P 6	464.37457	

$2\nu_3$ band of HCN occurring in the fourteenth order

Line J	First Order Wave Number (vac cm^{-1})	$\Delta\nu$	Line J	First Order Wave Number (vac cm^{-1})	$\Delta\nu$
P 6	464.37457		P 20	460.89638	
		0.22919	P 21	460.62594	0.27044
P 7	464.14538				0.27334
		0.23218	P 22	460.35260	
P 8	463.91320				0.27623
		0.23516	P 23	460.07637	
P 9	463.67804				0.27911
		0.23813	P 24	459.79726	
P 10	463.43991				0.28200
		0.24110	P 25	459.51526	
P 11	463.19881				0.28486
		0.24406	P 26	459.23040	
P 12	462.95475				0.28773
		0.24701	P 27	458.94267	
P 13	462.70774				0.29059
		0.24996	P 28	458.65208	
P 14	462.45778				0.29343
		0.25292	P 29	458.35865	
P 15	462.20486				0.29627
		0.25585	P 30	458.06238	
P 16	461.94901				0.29912
		0.25877	P 31	457.76326	
P 17	461.69024				0.30193
		0.26170	P 32	457.46133	
P 18	461.42854				0.30477
		0.26463	P 33	457.15656	
P 19	461.16391				
		0.26753			
P 20	460.89638				

First Order Wave Numbers of the Rotational Lines of the $2\nu_3$ Band of $HC^{12}N^{14}$

$2\nu_3$ band of HCN occurring in the fifteenth order

Line J	First Order Wave Number (vac cm^{-1})	$\Delta\nu$	Line J	First Order Wave Number (vac cm^{-1})	$\Delta\nu$
R 32	439.53944		R 13	437.10223	
		0.10206			0.15723
R 31	439.43738		R 12	436.94500	
		0.10499			0.16010
R 30	439.33239		R 11	436.78490	
		0.10790			0.16297
R 29	439.22449		R 10	436.62193	
		0.11084			0.16584
R 28	439.11365		R 9	436.45609	
		0.11374			0.16870
R 27	438.99991		R 8	436.28739	
		0.11668			0.17156
R 26	438.88323		R 7	436.11583	
		0.11958			0.17440
R 25	438.76365		R 6	435.94143	
		0.12250			0.17727
R 24	438.64115		R 5	435.76416	
		0.12540			0.18011
R 23	438.51575		R 4	435.58405	
		0.12832			0.18295
R 22	438.38743		R 3	435.40110	
		0.13122			0.18579
R 21	438.25621		R 2	435.21531	
		0.13412			0.18862
R 20	438.12209		R 1	435.02669	
		0.13703			0.19145
R 19	437.98506		R 0	434.83524	
		0.13991			0.39137
R 18	437.84515		P 1	434.44387	
		0.14281			0.19991
R 17	437.70234		P 2	434.24396	
		0.14570			0.20272
R 16	437.55664		P 3	434.04124	
		0.14859			0.20553
R 15	437.40805		P 4	433.83571	
		0.15147			0.20832
R 14	437.25658		P 5	433.62739	
		0.15435			0.21112
R 13	437.10223		P 6	433.41627	

$2\nu_3$ band of HCN occurring in the fifteenth order

Line J	First Order Wave Number (vac cm^{-1})	$\Delta\nu$	Line J	First Order Wave Number (vac cm^{-1})	$\Delta\nu$
P 6	433.41627		P 20	430.16995	
		0.21392			0.25241
P 7	433.20235		P 21	429.91754	
		0.21670			0.25511
P 8	432.98565		P 22	429.66243	
		0.21948			0.25782
P 9	432.76617		P 23	429.40461	
		0.22226			0.26050
P 10	432.54391		P 24	429.14411	
		0.22502			0.26320
P 11	432.31889		P 25	428.88091	
		0.22779			0.26587
P 12	432.09110		P 26	428.61504	
		0.23055			0.26855
P 13	431.86055		P 27	428.34649	
		0.23330			0.27122
P 14	431.62725		P 28	428.07527	
		0.23605			0.27386
P 15	431.39120		P 29	427.80141	
		0.23879			0.27652
P 16	431.15241		P 30	427.52489	
		0.24152			0.27918
P 17	430.91089		P 31	427.24571	
		0.24426			0.28180
P 18	430.66663		P 32	426.96391	
		0.24698			0.28445
P 19	430.41965		P 33	426.67946	
		0.24970			
P 20	430.16995				

First Order Wave Numbers of the Rotational Lines of the $2\nu_3$ Band of $HC^{12}N^{14}$

$2\nu_3$ band of HCN occurring in the sixteenth order

Line J	First Order Wave Number (vac cm^{-1})	$\Delta\nu$	Line J	First Order Wave Number (vac cm^{-1})	$\Delta\nu$
R 32	412.06823		R 13	409.78334	
		0.09569			0.14740
R 31	411.97254		R 12	409.63594	
		0.09842			0.15010
R 30	411.87412		R 11	409.48584	
		0.10116			0.15278
R 29	411.77296		R 10	409.33306	
		0.10391			0.15548
R 28	411.66905		R 9	409.17758	
		0.10664			0.15815
R 27	411.56241		R 8	409.01943	
		0.10938			0.16084
R 26	411.45303		R 7	408.85859	
		0.11210			0.16350
R 25	411.34093		R 6	408.69509	
		0.11485			0.16619
R 24	411.22608		R 5	408.52890	
		0.11757			0.16885
R 23	411.10851		R 4	408.36005	
		0.12029			0.17152
R 22	410.98822		R 3	408.18853	
		0.12303			0.17417
R 21	410.86519		R 2	408.01436	
		0.12573			0.17683
R 20	410.73946		R 1	407.83753	
		0.12847			0.17949
R 19	410.61099		R 0	407.65804	
		0.13116			0.36691
R 18	410.47983		P 1	407.29113	
		0.13389			0.18742
R 17	410.34594		P 2	407.10371	
		0.13659			0.19005
R 16	410.20935		P 3	406.91366	
		0.13931			0.19268
R 15	410.07004		P 4	406.72098	
		0.14200			0.19530
R 14	409.92804		P 5	406.52568	
		0.14470			0.19793
R 13	409.78334		P 6	406.32775	

$2\nu_3$ band of HCN occurring in the sixteenth order

Line J	First Order Wave Number (vac cm^{-1})	$\Delta\nu$	Line J	First Order Wave Number (vac cm^{-1})	$\Delta\nu$
P 6	406.32775		P 20	403.28433	
		0.20054			0.23664
P 7	406.12721		P 21	403.04769	
		0.20316			0.23916
P 8	405.92405		P 22	402.80853	
		0.20577			0.24170
P 9	405.71828		P 23	402.56683	
		0.20836			0.24423
P 10	405.50992		P 24	402.32260	
		0.21096			0.24675
P 11	405.29896		P 25	402.07585	
		0.21355			0.24925
P 12	405.08541		P 26	401.82660	
		0.21614			0.25176
P 13	404.86927		P 27	401.57484	
		0.21872			0.25427
P 14	404.65055		P 28	401.32057	
		0.22130			0.25675
P 15	404.42925		P 29	401.06382	
		0.22386			0.25924
P 16	404.20539		P 30	400.80458	
		0.22643			0.26172
P 17	403.97896		P 31	400.54286	
		0.22899			0.26420
P 18	403.74997		P 32	400.27866	
		0.23154			0.26667
P 19	403.51843		P 33	400.01199	
		0.23410			
P 20	403.28433				

First Order Wave Numbers of the Rotational Lines of the $(3\nu_1 + \nu_3)$ Band of $N_2^{14}O^{16}$

$3\nu_1 + \nu_3$ band of N_2O occurring in the twelfth order

Line J	First Order Wave Number (vac cm^{-1})	$\Delta\nu$	Line J	First Order Wave Number (vac cm^{-1})	$\Delta\nu$
R 40	499.57643		R 20	499.04967	
		0.01350			0.04068
R 39	499.56293		R 19	499.00899	
		0.01483			0.04207
R 38	499.54810		R 18	498.96692	
		0.01618			0.04345
R 37	499.53192		R 17	498.92347	
		0.01751			0.04482
R 36	499.51441		R 16	498.87865	
		0.01886			0.04622
R 35	499.49555		R 15	498.83243	
		0.02021			0.04760
R 34	499.47534		R 14	498.78483	
		0.02156			0.04898
R 33	499.45378		R 13	498.73585	
		0.02291			0.05037
R 32	499.43087		R 12	498.68548	
		0.02425			0.05176
R 31	499.40662		R 11	498.63372	
		0.02563			0.05315
R 30	499.38099		R 10	498.58057	
		0.02698			0.05454
R 29	499.35401		R 9	498.52603	
		0.02834			0.05591
R 28	499.32567		R 8	498.47012	
		0.02970			0.05732
R 27	499.29597		R 7	498.41280	
		0.03107			0.05871
R 26	499.26490		R 6	498.35409	
		0.03244			0.06010
R 25	499.23246		R 5	498.29399	
		0.03381			0.06148
R 24	499.19865		R 4	498.23251	
		0.03518			0.06289
R 23	499.16347		R 3	498.16962	
		0.03656			0.06426
R 22	499.12691		R 2	498.10536	
		0.03793			0.06567
R 21	499.08898		R 1	498.03969	
		0.03931			0.06705
R 20	499.04967		R 0	497.97264	

$3\nu_1 + \nu_3$ band of N_2O occurring in the twelfth order

Line J	First Order Wave Number (vac cm^{-1})	$\Delta\nu$	Line J	First Order Wave Number (vac cm^{-1})	$\Delta\nu$
R 0	497.97264		P 21	496.14676	
		0.13828			0.09884
P 1	497.83436		P 22	496.04792	
		0.07123			0.10019
P 2	497.76313		P 23	495.94773	
		0.07261			0.10155
P 3	497.69052		P 24	495.84618	
		0.07400			0.10291
P 4	497.61652		P 25	495.74327	
		0.07540			0.10427
P 5	497.54112		P 26	495.63900	
		0.07678			0.10563
P 6	497.46434		P 27	495.53337	
		0.07817			0.10697
P 7	497.38617		P 28	495.42640	
		0.07955			0.10833
P 8	497.30662		P 29	495.31807	
		0.08095			0.10967
P 9	497.22567		P 30	495.20840	
		0.08232			0.11101
P 10	497.14335		P 31	495.09739	
		0.08372			0.11236
P 11	497.05963		P 32	494.98503	
		0.08509			0.11369
P 12	496.97454		P 33	494.87134	
		0.08647			0.11502
P 13	496.88807		P 34	494.75632	
		0.08785			0.11636
P 14	496.80022		P 35	494.63996	
		0.08923			0.11769
P 15	496.71099		P 36	494.52227	
		0.09061			0.11900
P 16	496.62038		P 37	494.40327	
		0.09198			0.12034
P 17	496.52840		P 38	494.28293	
		0.09336			0.12165
P 18	496.43504		P 39	494.16128	
		0.09472			0.12296
P 19	496.34032		P 40	494.03832	
		0.09610			
P 20	496.24422				
		0.09746			
P 21	496.14676				

First Order Wave Numbers of the Rotational Lines of the $(3\nu_1 + \nu_3)$ Band of $N_2^{14}O^{16}$

$3\nu_1 + \nu_3$ band of N_2O occurring in the thirteenth order

Line J	First Order Wave Number (vac cm^{-1})	$\Delta\nu$	Line J	First Order Wave Number (vac cm^{-1})	$\Delta\nu$
R 40	461.14748		R 20	460.66124	
		0.01246			0.03756
R 39	461.13502		R 19	460.62368	
		0.01370			0.03883
R 38	461.12132		R 18	460.58485	
		0.01493			0.04010
R 37	461.10639		R 17	460.54475	
		0.01617			0.04138
R 36	461.09022		R 16	460.50337	
		0.01740			0.04266
R 35	461.07282		R 15	460.46071	
		0.01866			0.04394
R 34	461.05416		R 14	460.41677	
		0.01990			0.04522
R 33	461.03426		R 13	460.37155	
		0.02114			0.04649
R 32	461.01312		R 12	460.32506	
		0.02240			0.04778
R 31	460.99072		R 11	460.27728	
		0.02365			0.04906
R 30	460.96707		R 10	460.22822	
		0.02491			0.05034
R 29	460.94216		R 9	460.17788	
		0.02615			0.05162
R 28	460.91601		R 8	460.12626	
		0.02743			0.05291
R 27	460.88858		R 7	460.07335	
		0.02867			0.05419
R 26	460.85991		R 6	460.01916	
		0.02995			0.05548
R 25	460.82996		R 5	459.96368	
		0.03121			0.05675
R 24	460.79875		R 4	459.90693	
		0.03247			0.05805
R 23	460.76628		R 3	459.84888	
		0.03375			0.05932
R 22	460.73253		R 2	459.78956	
		0.03501			0.06061
R 21	460.69752		R 1	459.72895	
		0.03628			0.06190
R 20	460.66124		R 0	459.66705	

$3\nu_1 + \nu_3$ band of N_2O occurring in the thirteenth order

Line J	First Order Wave Number (vac cm^{-1})	$\Delta\nu$	Line J	First Order Wave Number (vac cm^{-1})	$\Delta\nu$
R 0	459.66705		P 20	458.07158	
		0.12764			0.08996
P 1	459.53941		P 21	457.98162	
		0.06575			0.09123
P 2	459.47366		P 22	457.89039	
		0.06703			0.09248
P 3	459.40663		P 23	457.79791	
		0.06831			0.09374
P 4	459.33832		P 24	457.70417	
		0.06959			0.09500
P 5	459.26873		P 25	457.60917	
		0.07088			0.09625
P 6	459.19785		P 26	457.51292	
		0.07216			0.09750
P 7	459.12569		P 27	457.41542	
		0.07343			0.09874
P 8	459.05226		P 28	457.31668	
		0.07471			0.10000
P 9	458.97755		P 29	457.21668	
		0.07600			0.10123
P 10	458.90155		P 30	457.11545	
		0.07727			0.10247
P 11	458.82428		P 31	457.01298	
		0.07855			0.10372
P 12	458.74573		P 32	456.90926	
		0.07981			0.10494
P 13	458.66592		P 33	456.80432	
		0.08110			0.10618
P 14	458.58482		P 34	456.69814	
		0.08237			0.10741
P 15	458.50245		P 35	456.59073	
		0.08363			0.10863
P 16	458.41882		P 36	456.48210	
		0.08491			0.10985
P 17	458.33391		P 37	456.37225	
		0.08618			0.11108
P 18	458.24773		P 38	456.26117	
		0.08744			0.11229
P 19	458.16029		P 39	456.14888	
		0.08871			0.11351
P 20	458.07158		P 40	456.03537	

First Order Wave Numbers of the Rotational Lines of the $(3\nu_1 + \nu_3)$ Band of $N_2^{14}O^{16}$

$3\nu_1 + \nu_3$ band of N_2O occurring in the fourteenth order

Line J	First Order Wave Number (vac cm^{-1})	$\Delta\nu$	Line J	First Order Wave Number (vac cm^{-1})	$\Delta\nu$
R 40	428.20837		R 20	427.75686	
		0.01157			0.03487
R 39	428.19680		R 19	427.72199	
		0.01271			0.03605
R 38	428.18409		R 18	427.68594	
		0.01387			0.03725
R 37	428.17022		R 17	427.64869	
		0.01501			0.03842
R 36	428.15521		R 16	427.61027	
		0.01617			0.03961
R 35	428.13904		R 15	427.57066	
		0.01732			0.04080
R 34	428.12172		R 14	427.52986	
		0.01848			0.04199
R 33	428.10324		R 13	427.48787	
		0.01963			0.04317
R 32	428.08361		R 12	427.44470	
		0.02080			0.04437
R 31	428.06281		R 11	427.40033	
		0.02196			0.04555
R 30	428.04085		R 10	427.35478	
		0.02313			0.04675
R 29	428.01772		R 9	427.30803	
		0.02428			0.04793
R 28	427.99344		R 8	427.26010	
		0.02547			0.04913
R 27	427.96797		R 7	427.21097	
		0.02663			0.05032
R 26	427.94134		R 6	427.16065	
		0.02780			0.05151
R 25	427.91354		R 5	427.10914	
		0.02898			0.05270
R 24	427.88456		R 4	427.05644	
		0.03016			0.05390
R 23	427.85440		R 3	427.00254	
		0.03134			0.05509
R 22	427.82306		R 2	426.94745	
		0.03250			0.05629
R 21	427.79056		R 1	426.89116	
		0.03370			0.05747
R 20	427.75686		R 0	426.83369	

$3\nu_1 + \nu_3$ band of N$_2$O occurring in the fourteenth order

Line J	First Order Wave Number (vac cm^{-1})	$\Delta\nu$	Line J	First Order Wave Number (vac cm^{-1})	$\Delta\nu$
R 0	426.83369		P 20	425.35219	
		0.11853			0.08354
P 1	426.71516		P 21	425.26865	
		0.06105			0.08471
P 2	426.65411		P 22	425.18394	
		0.06224			0.08588
P 3	426.59187		P 23	425.09806	
		0.06343			0.08705
P 4	426.52844		P 24	425.01101	
		0.06462			0.08821
P 5	426.46382		P 25	424.92280	
		0.06581			0.08937
P 6	426.39801		P 26	424.83343	
		0.06701			0.09054
P 7	426.33100		P 27	424.74289	
		0.06819			0.09169
P 8	426.26281		P 28	424.65120	
		0.06937			0.09285
P 9	426.19344		P 29	424.55835	
		0.07057			0.09401
P 10	426.12287		P 30	424.46434	
		0.07176			0.09515
P 11	426.05111		P 31	424.36919	
		0.07293			0.09630
P 12	425.97818		P 32	424.27289	
		0.07412			0.09745
P 13	425.90406		P 33	424.17544	
		0.07530			0.09860
P 14	425.82876		P 34	424.07684	
		0.07648			0.09973
P 15	425.75228		P 35	423.97711	
		0.07767			0.10087
P 16	425.67461		P 36	423.87624	
		0.07884			0.10201
P 17	425.59577		P 37	423.77423	
		0.08002			0.10314
P 18	425.51575		P 38	423.67109	
		0.08119			0.10428
P 19	425.43456		P 39	423.56681	
		0.08237			0.10540
P 20	425.35219		P 40	423.46141	

APPENDIX III. 16

First Order Wave Numbers of the Rotational Lines of the $(\nu_1 + \nu_3)$ Band of HC^{12}N^{14}

$\nu_1 + \nu_3$ band of HCN occurring in the eleventh order

Line J	First Order Wave Number (vac cm^{-1})	$\Delta\nu$	Line J	First Order Wave Number (vac cm^{-1})	$\Delta\nu$
R 28	496.50118		R 9	492.82000	
		0.15936			0.23164
R 27	496.34182		R 8	492.58836	
		0.16318			0.23536
R 26	496.17864		R 7	492.35300	
		0.16709			0.23909
R 25	496.01155		R 6	492.11391	
		0.17091			0.24282
R 24	495.84064		R 5	491.87109	
		0.17473			0.24654
R 23	495.66591		R 4	491.62455	
		0.17855			0.25019
R 22	495.48736		R 3	491.37436	
		0.18236			0.25400
R 21	495.30500		R 2	491.12036	
		0.18618			0.25772
R 20	495.11882		R 1	490.86264	
		0.19009			0.26137
R 19	494.92873		R 0	490.60127	
		0.19382			0.53382
R 18	494.73491		P 1	490.06745	
		0.19764			0.27245
R 17	494.53727		P 2	489.79500	
		0.2ᴗ145			0.27609
R 16	494.33582		P 3	489.51891	
		0.20518			0.27982
R 15	494.13064		P 4	489.23909	
		0.20900			0.28345
R 14	493.92164		P 5	488.95564	
		0.21282			0.28709
R 13	493.70882		P 6	488.66855	
		0.21655			0.29082
R 12	493.49227		P 7	488.37773	
		0.22036			0.29437
R 11	493.27191		P 8	488.08336	
		0.22409			0.29809
R 10	493.04782		P 9	487.78527	
		0.22782			0.30163
R 9	492.82000		P 10	487.48364	

$v_1 + v_3$ band of HCN occurring in the eleventh order

Line J	First Order Wave Number (vac cm^{-1})	Δv	Line J	First Order Wave Number (vac cm^{-1})	Δv
P 10	487.48364		P 19	484.60636	
		0.30528			0.33763
P 11	487.17836		P 20	484.26873	
		0.30891			0.34118
P 12	486.86945		P 21	483.92755	
		0.31254			0.34464
P 13	486.55691		P 22	483.58291	
		0.31609			0.34827
P 14	486.24082		P 23	483.23464	
		0.31973			0.35182
P 15	485.92109		P 24	482.88282	
		0.32336			0.35527
P 16	485.59773		P 25	482.52755	
		0.32691			0.35882
P 17	485.27082		P 26	482.16873	
		0.33046			0.36237
P 18	484.94036		P 27	481.80636	
		0.33400			0.36581
P 19	484.60636		P 28	481.44055	

First Order Wave Numbers of the Rotational Lines of the $(\nu_1 + \nu_3)$ Band of the $HC^{12}N^{14}$

$\nu_1 + \nu_3$ band of HCN occurring in the twelfth order

Line J	First Order Wave Number (vac cm^{-1})	$\Delta\nu$	Line J	First Order Wave Number (vac cm^{-1})	$\Delta\nu$
R 28	455.12608		R 9	451.75167	
		0.14608			0.21234
R 27	454.98000		R 8	451.53933	
		0.14958			0.21575
R 26	454.83042		R 7	451.32358	
		0.15317			0.21916
R 25	454.67725		R 6	451.10442	
		0.15667			0.22259
R 24	454.52058		R 5	450.88183	
		0.16016			0.22600
R 23	454.36042		R 4	450.65583	
		0.16367			0.22933
R 22	454.19675		R 3	450.42650	
		0.16717			0.23283
R 21	454.02958		R 2	450.19367	
		0.17066			0.23625
R 20	453.85892		R 1	449.95742	
		0.17425			0.23959
R 19	453.68467		R 0	449.71783	
		0.17767			0.48933
R 18	453.50700		P 1	449.22850	
		0.18117			0.24975
R 17	453.32583		P 2	448.97875	
		0.18466			0.25308
R 16	453.14117		P 3	448.72567	
		0.18809			0.25650
R 15	452.95308		P 4	448.46917	
		0.19158			0.25984
R 14	452.76150		P 5	448.20933	
		0.19508			0.26316
R 13	452.56642		P 6	447.94617	
		0.19850			0.26659
R 12	452.36792		P 7	447.67958	
		0.20200			0.26983
R 11	452.16592		P 8	447.40975	
		0.20542			0.27325
R 10	451.96050		P 9	447.13650	
		0.20883			0.27650
R 9	451.75167		P 10	446.86000	

$v_1 + v_3$ band of HCN occurring in the twelfth order

Line J	First Order Wave Number (vac cm^{-1})	Δv	Line J	First Order Wave Number (vac cm^{-1})	Δv
P 10	446.86000		P 19	444.22250	
		0.27983			0.30950
P 11	446.58017		P 20	443.91300	
		0.28317			0.31275
P 12	446.29700		P 21	443.60025	
		0.28650			0.31592
P 13	446.01050		P 22	443.28433	
		0.28975			0.31925
P 14	445.72075		P 23	442.96508	
		0.29308			0.32250
P 15	445.42767		P 24	442.64258	
		0.29642			0.32566
P 16	445.13125		P 25	442.31692	
		0.29967			0.32892
P 17	444.83158		P 26	441.98800	
		0.30291			0.33217
P 18	444.52867		P 27	441.65583	
		0.30617			0.33533
P 19	444.22250		P 28	441.32050	

First Order Wave Numbers of the Rotational Lines of the $(\nu_1 + \nu_3)$ Band of $HC^{12}N^{14}$

$\nu_1 + \nu_3$ band of HCN occurring in the thirteenth order

Line J	First Order Wave Number (vac cm^{-1})	$\Delta\nu$	Line J	First Order Wave Number (vac cm^{-1})	$\Delta\nu$
R 28	420.11638		R 9	417.00154	
		0.13484			0.19600
R 27	419.98154		R 8	416.80554	
		0.13808			0.19916
R 26	419.84346		R 7	416.60638	
		0.14138			0.20230
R 25	419.70208		R 6	416.40408	
		0.14462			0.20546
R 24	419.55746		R 5	416.19862	
		0.14784			0.20862
R 23	419.40962		R 4	415.99000	
		0.15108			0.21169
R 22	419.25854		R 3	415.77831	
		0.15431			0.21493
R 21	419.10423		R 2	415.56338	
		0.15754			0.21807
R 20	418.94669		R 1	415.34531	
		0.16084			0.22116
R 19	418.78585		R 0	415.12415	
		0.16400			0.45169
R 18	418.62185		P 1	414.67246	
		0.16723			0.23054
R 17	418.45462		P 2	414.44192	
		0.17047			0.23361
R 16	418.28415		P 3	414.20831	
		0.17361			0.23677
R 15	418.11054		P 4	413.97154	
		0.17685			0.23985
R 14	417.93369		P 5	413.73169	
		0.18007			0.24292
R 13	417.75362		P 6	413.48877	
		0.18324			0.24608
R 12	417.57038		P 7	413.24269	
		0.18646			0.24907
R 11	417.38392		P 8	412.99362	
		0.18961			0.25224
R 10	417.19431		P 9	412.74138	
		0.19277			0.25523
R 9	417.00154		P 10	412.48615	

$v_1 + v_3$ band of HCN occurring in the thirteenth order

Line J	First Order Wave Number (vac cm^{-1})	Δv	Line J	First Order Wave Number (vac cm^{-1})	Δv
P 10	412.48615		P 19	410.05154	
		0.25830			0.28569
P 11	412.22785		P 20	409.76585	
		0.26139			0.28870
P 12	411.96646		P 21	409.47715	
		0.26446			0.29161
P 13	411.70200		P 22	409.18554	
		0.26746			0.29469
P 14	411.43454		P 23	408.89085	
		0.27054			0.29770
P 15	411.16400		P 24	408.59315	
		0.27362			0.30061
P 16	410.89038		P 25	408.29254	
		0.27661			0.30362
P 17	410.61377		P 26	407.98892	
		0.27962			0.30661
P 18	410.33415		P 27	407.68231	
		0.28261			0.30954
P 19	410.05154		P 28	407.37277	

First Order Wave Numbers of the Rotational Lines of the $(2\nu_1 + \nu_3)$ Band of $N_2^{14}O^{16}$

$2\nu_1 + \nu_3$ band of $N_2^{14}O^{16}$ occurring in the eleventh order

Line J	First Order Wave Number (vac cm^{-1})	$\Delta\nu$	Line J	First Order Wave Number (vac cm^{-1})	$\Delta\nu$
R 54	432.33843		R 35	431.98268	
		0.00767			0.03109
R 53	432.33076		R 34	431.95159	
		0.00888			0.03234
R 52	432.32188		R 33	431.91925	
		0.01010			0.03359
R 51	432.31178		R 32	431.88566	
		0.01133			0.03482
R 50	432.30045		R 31	431.85084	
		0.01255			0.03608
R 49	432.28790		R 30	431.81476	
		0.01378			0.03732
R 48	432.27412		R 29	431.77744	
		0.01501			0.03858
R 47	432.25911		R 28	431.73886	
		0.01625			0.03982
R 46	432.24286		R 27	431.69904	
		0.01747			0.04108
R 45	432.22539		R 26	431.65796	
		0.01870			0.04231
R 44	432.20669		R 25	431.61565	
		0.01994			0.04358
R 43	432.18675		R 24	431.57207	
		0.02117			0.04482
R 42	432.16558		R 23	431.52725	
		0.02242			0.04608
R 41	432.14316		R 22	431.48117	
		0.02364			0.04733
R 40	432.11952		R 21	431.43384	
		0.02488			0.04859
R 39	432.09464		R 20	431.38525	
		0.02613			0.04982
R 38	432.06851		R 19	431.33543	
		0.02736			0.05109
R 37	432.04115		R 18	431.28434	
		0.02861			0.05235
R 36	432.01254		R 17	431.23199	
		0.02986			0.05360
R 35	431.98268		R 16	431.17839	

$2\nu_1 + \nu_3$ band of $N_2^{14}O^{16}$ occurring in the eleventh order

Line J	First Order Wave Number (vac cm^{-1})	$\Delta\nu$	Line J	First Order Wave Number (vac cm^{-1})	$\Delta\nu$
R 16	431.17839		P 5	429.68178	
		0.05484			0.08244
R 15	431.12355		P 6	429.59934	
		0.05611			0.08370
R 14	431.06744		P 7	429.51564	
		0.05737			0.08495
R 13	431.01007		P 8	429.43069	
		0.05861			0.08619
R 12	430.95146		P 9	429.34450	
		0.05988			0.08745
R 11	430.89158		P 10	429.25705	
		0.06113			0.08869
R 10	430.83045		P 11	429.16836	
		0.06238			0.08993
R 9	430.76807		P 12	429.07843	
		0.06363			0.09119
R 8	430.70444		P 13	428.98724	
		0.06489			0.09243
R 7	430.63955		P 14	428.89481	
		0.06616			0.09367
R 6	430.57339		P 15	428.80114	
		0.06740			0.09492
R 5	430.50599		P 16	428.70622	
		0.06865			0.09617
R 4	430.43734		P 17	428.61005	
		0.06992			0.09740
R 3	430.36742		P 18	428.51265	
		0.07117			0.09865
R 2	430.29625		P 19	428.41400	
		0.07242			0.09989
R 1	430.22383		P 20	428.31411	
		0.07368			0.10112
R 0	430.15015		P 21	428.21299	
		0.15111			0.10236
P 1	429.99904		P 22	428.11063	
		0.07744			0.10360
P 2	429.92160		P 23	428.00703	
		0.07869			0.10484
P 3	429.84291		P 24	427.90219	
		0.07994			0.10607
P 4	429.76297		P 25	427.79612	
		0.08119			0.10730
P 5	429.68178		P 26	427.68882	

$2\nu_1 + \nu_3$ band of $N_2^{14}O^{16}$ occurring in the eleventh order

Line J	First Order Wave Number (vac cm^{-1})	$\Delta\nu$	Line J	First Order Wave Number (vac cm^{-1})	$\Delta\nu$
P 26	427.68882		P 41	425.93222	
		0.10854			0.12685
P 27	427.58028		P 42	425.80537	
		0.10977			0.12804
P 28	427.47051		P 43	425.67733	
		0.11099			0.12925
P 29	427.35952		P 44	425.54808	
		0.11223			0.13045
P 30	427.24729		P 45	425.41763	
		0.11344			0.13166
P 31	427.13385		P 46	425.28597	
		0.11469			0.13285
P 32	427.01916		P 47	425.15312	
		0.11589			0.13406
P 33	426.90327		P 48	425.01906	
		0.11712			0.13524
P 34	426.78615		P 49	424.88382	
		0.11834			0.13644
P 35	426.66781		P 50	424.74738	
		0.11956			0.13763
P 36	426.54825		P 51	424.60975	
		0.12078			0.13882
P 37	426.42747		P 52	424.47093	
		0.12200			0.14001
P 38	426.30547		P 53	424.33092	
		0.12321			0.14119
P 39	426.18226		P 54	424.18973	
		0.12441			0.14238
P 40	426.05785		P 55	424.04735	
		0.12563			
P 41	425.93222				

First Order Wave Numbers of the Rotational Lines of the $(\nu_1 + 2\nu_2 + \nu_3)$ Band of $N_2^{14}O^{16}$

$\nu_1 + 2\nu_2 + \nu_3$ band of $N_2^{14}O^{16}$ occurring in the eleventh order

Line J	First Order Wave Number (vac cm^{-1})	$\Delta\nu$	Line J	First Order Wave Number (vac cm^{-1})	$\Delta\nu$
R 48	423.67095		R 29	422.84356	
		0.03542			0.05224
R 47	423.63553		R 28	422.79132	
		0.03637			0.05309
R 46	423.59916		R 27	422.73823	
		0.03729			0.05391
R 45	423.56187		R 26	422.68432	
		0.03821			0.05476
R 44	423.52366		R 25	422.62956	
		0.03912			0.05557
R 43	423.48454		R 24	422.57399	
		0.04004			0.05639
R 42	423.44450		R 23	422.51760	
		0.04095			0.05722
R 41	423.40355		R 22	422.46038	
		0.04183			0.05803
R 40	423.36172		R 21	422.40235	
		0.04274			0.05885
R 39	423.31898		R 20	422.34350	
		0.04363			0.05965
R 38	423.27535		R 19	422.28385	
		0.04450			0.06047
R 37	423.23085		R 18	422.22338	
		0.04538			0.06126
R 36	423.18547		R 17	422.16212	
		0.04627			0.06207
R 35	423.13920		R 16	422.10005	
		0.04713			0.06288
R 34	423.09207		R 15	422.03717	
		0.04800			0.06366
R 33	423.04407		R 14	421.97351	
		0.04884			0.06446
R 32	422.99523		R 13	421.90905	
		0.04971			0.06526
R 31	422.94552		R 12	421.84379	
		0.05056			0.06604
R 30	422.89496		R 11	421.77775	
		0.05140			0.06683
R 29	422.84356		R 10	421.71092	

$\nu_1 + 2\nu_2 + \nu_3$ band of $N_2^{14}O^{16}$ occurring in the eleventh order

Line J	First Order Wave Number (vac cm^{-1})	$\Delta\nu$	Line J	First Order Wave Number (vac cm^{-1})	$\Delta\nu$
R 10	421.71092		P 9	420.21066	
		0.06762			0.08314
R 9	421.64330		P 10	420.12752	
		0.06840			0.08393
R 8	421.57490		P 11	420.04359	
		0.06919			0.08471
R 7	421.50571		P 12	419.95888	
		0.06997			0.08547
R 6	421.43574		P 13	419.87341	
		0.07075			0.08626
R 5	421.36499		P 14	419.78715	
		0.07153			0.08704
R 4	421.29346		P 15	419.70011	
		0.07230			0.08782
R 3	421.22116		P 16	419.61229	
		0.07309			0.08861
R 2	421.14807		P 17	419.52368	
		0.07385			0.08938
R 1	421.07422		P 18	419.43430	
		0.07464			0.09017
R 0	420.99958		P 19	419.34413	
		0.15159			0.09097
P 1	420.84799		P 20	419.25316	
		0.07695			0.09175
P 2	420.77104		P 21	419.16141	
		0.07774			0.09254
P 3	420.69330		P 22	419.06887	
		0.07850			0.09333
P 4	420.61480		P 23	418.97554	
		0.07928			0.09414
P 5	420.53552		P 24	418.88140	
		0.08006			0.09493
P 6	420.45546		P 25	418.78647	
		0.08082			0.09573
P 7	420.37464		P 26	418.69074	
		0.08160			0.09654
P 8	420.29304		P 27	418.59420	
		0.08238			0.09734
P 9	420.21066		P 28	418.49686	

$v_1 + 2v_2 + v_3$ band of $N_2^{14}O^{16}$ occurring in the eleventh order

Line J	First Order Wave Number (vac cm^{-1})	Δv	Line J	First Order Wave Number (vac cm^{-1})	Δv
P 28	418.49686		P 38	417.47840	
		0.09815			0.10644
P 29	418.39871		P 39	417.37196	
		0.09896			0.10727
P 30	418.29975		P 40	417.26469	
		0.09979			0.10812
P 31	418.19996		P 41	417.15657	
		0.10059			0.10900
P 32	418.09937		P 42	417.04757	
		0.10142			0.10985
P 33	417.99795		P 43	416.93772	
		0.10224			0.11072
P 34	417.89571		P 44	416.82700	
		0.10308			0.11159
P 35	417.79263		P 45	416.71541	
		0.10390			0.11246
P 36	417.68873		P 46	416.60295	
		0.10475			0.11335
P 37	417.58398		P 47	416.48960	
		0.10558			
P 38	417.47840				

First Order Wave Numbers of the Rotational Lines of the 2-0 Band of $C^{12}O^{16}$

2-0 band of CO occurring in the ninth order

Line J	First Order Wave Number (vac cm^{-1})	$\Delta \nu$	Line J	First Order Wave Number (vac cm^{-1})	$\Delta \nu$
R 30	482.64512		R 11	477.85576	
		0.17850			0.33282
R 29	482.46662		R 10	477.52294	
		0.18676			0.34075
R 28	482.27986		R 9	477.18219	
		0.19503			0.34870
R 27	482.08483		R 8	476.83349	
		0.20325			0.35662
R 26	481.88158		R 7	476.47687	
		0.21147			0.36453
R 25	481.67011		R 6	476.11234	
		0.21967			0.37243
R 24	481.45044		R 5	475.73991	
		0.22785			0.38030
R 23	481.22259		R 4	475.35961	
		0.23603			0.38817
R 22	480.98656		R 3	474.97144	
		0.24418			0.39600
R 21	480.74238		R 2	474.57544	
		0.25231			0.40383
R 20	480.49007		R 1	474.17161	
		0.26045			0.41165
R 19	480.22962		R 0	473.75996	
		0.26853			0.84667
R 18	479.96109		P 1	472.91329	
		0.27663			0.43499
R 17	479.68446		P 2	472.47830	
		0.28470			0.44273
R 16	479.39976		P 3	472.03557	
		0.29276			0.45047
R 15	479.10700		P 4	471.58510	
		0.30080			0.45818
R 14	478.80620		P 5	471.12692	
		0.30882			0.46588
R 13	478.49738		P 6	470.66104	
		0.31682			0.47355
R 12	478.18056		P 7	470.18749	
		0.32480			0.48122
R 11	477.85576		P 8	469.70627	

2-0 band of CO occurring in the ninth order

Line J	First Order Wave Number (vac cm⁻¹)	$\Delta\nu$	Line J	First Order Wave Number (vac cm⁻¹)	$\Delta\nu$
P 8	469.70627		P 19	463.91157	
		0.48888			0.57193
P 9	469.21739		P 20	463.33964	
		0.49650			0.57938
P 10	468.72089		P 21	462.76026	
		0.50412			0.58683
P 11	468.21677		P 22	462.17343	
		0.51173			0.59423
P 12	467.70504		P 23	461.57920	
		0.51931			0.60164
P 13	467.18573		P 24	460.97756	
		0.52686			0.60904
P 14	466.65887		P 25	460.36852	
		0.53443			0.61640
P 15	466.12444		P 26	459.75212	
		0.54195			0.62375
P 16	465.58249		P 27	459.12837	
		0.54948			0.63109
P 17	465.03301		P 28	458.49728	
		0.55698			0.63842
P 18	464.47603		P 29	457.85886	
		0.56446			0.64572
P 19	463.91157		P 30	457.21314	

First Order Wave Numbers of the Rotational Lines of the ν_3 Band of $HC^{12}N^{14}$

ν_3 band of HCN occurring in the seventh order

Line J	First Order Wave Number (vac cm^{-1})	$\Delta\nu$	Line J	First Order Wave Number (vac cm^{-1})	$\Delta\nu$
R 25	482.97666		R 6	475.94074	
		0.34195			0.40130
R 24	482.63471		R 5	475.53944	
		0.34513			0.40434
R 23	482.28958		R 4	475.13510	
		0.34825			0.40736
R 22	481.94133		R 3	474.72774	
		0.35157			0.41037
R 21	481.58976		R 2	474.31737	
		0.35469			0.41338
R 20	481.23507		R 1	473.90399	
		0.35786			0.41639
R 19	480.87721		R 0	473.48760	
		0.36101			0.84171
R 18	480.51620		P 1	472.64589	
		0.36416			0.42532
R 17	480.15204		P 2	472.22057	
		0.36731			0.42827
R 16	479.78473		P 3	471.79230	
		0.37043			0.43123
R 15	479.41430		P 4	471.36107	
		0.37357			0.43414
R 14	479.04073		P 5	470.92693	
		0.37669			0.43709
R 13	478.66404		P 6	470.48984	
		0.37978			0.44000
R 12	478.28426		P 7	470.04984	
		0.38289			0.44291
R 11	477.90137		P 8	469.60693	
		0.38598			0.44580
R 10	477.51539		P 9	469.16113	
		0.38906			0.44869
R 9	477.12633		P 10	468.71244	
		0.39214			0.45155
R 8	476.73419		P 11	468.26089	
		0.39520			0.45442
R 7	476.33899		P 12	467.80647	
		0.39825			0.45728
R 6	475.94074		P 13	467.34919	

ν_3 band of HCN occurring in the seventh order

Line J	First Order Wave Number (vac cm^{-1})	$\Delta\nu$	Line J	First Order Wave Number (vac cm^{-1})	$\Delta\nu$
P 13	467.34919		P 20	464.06933	
		0.46012			0.47970
P 14	466.88907		P 21	463.58963	
		0.46294			0.48244
P 15	466.42613		P 22	463.10719	
		0.46577			0.48526
P 16	465.96036		P 23	462.62193	
		0.46863			0.48784
P 17	465.49173		P 24	462.13409	
		0.47132			0.49065
P 18	465.02041		P 25	461.64344	
		0.47415			0.49333
P 19	464.54626		P 26	461.15011	
		0.47693			0.49602
P 20	464.06933		P 27	460.65409	

First Order Wave Numbers of the Rotational Lines of the ν_3 Band of $HC^{12}N^{14}$

ν_3 band of HCN occurring in the eighth order

Line J	First Order Wave Number (vac cm^{-1})	$\Delta\nu$	Line J	First Order Wave Number (vac cm^{-1})	$\Delta\nu$
R 25	422.60458		R 6	416.44815	
		0.29920			0.35114
R 24	422.30538		R 5	416.09701	
		0.30199			0.35380
R 23	422.00339		R 4	415.74321	
		0.30473			0.35643
R 22	421.69866		R 3	415.38678	
		0.30762			0.35908
R 21	421.39104		R 2	415.02770	
		0.31035			0.36171
R 20	421.08069		R 1	414.66599	
		0.31313			0.36434
R 19	420.76756		R 0	414.30165	
		0.31588			0.73650
R 18	420.45168		P 1	413.56515	
		0.31864			0.37215
R 17	420.13304		P 2	413.19300	
		0.32140			0.37474
R 16	419.81164		P 3	412.81826	
		0.32413			0.37732
R 15	419.48751		P 4	412.44094	
		0.32687			0.37988
R 14	419.16064		P 5	412.06106	
		0.32960			0.38245
R 13	418.83104		P 6	411.67861	
		0.33231			0.38500
R 12	418.49873		P 7	411.29361	
		0.33503			0.38755
R 11	418.16370		P 8	410.90606	
		0.33774			0.39007
R 10	417.82596		P 9	410.51599	
		0.34042			0.39260
R 9	417.48554		P 10	410.12339	
		0.34313			0.39511
R 8	417.14241		P 11	409.72828	
		0.34580			0.39762
R 7	416.79661		P 12	409.33066	
		0.34846			0.40012
R 6	416.44815		P 13	408.93054	

ν_3 band of HCN occurring in the eighth order

Line J	First Order Wave Number (vac cm^{-1})	$\Delta\nu$	Line J	First Order Wave Number (vac. cm^{-1})	$\Delta\nu$
P 13	408.93054		P 20	406.06066	
		0.40260			0.41973
P 14	408.52794		P 21	405.64093	
		0.40508			0.42214
P 15	408.12286		P 22	405.21879	
		0.40755			0.42460
P 16	407.71531		P 23	404.79419	
		0.41005			0.42686
P 17	407.30526		P 24	404.36733	
		0.41240			0.42932
P 18	406.89286		P 25	403.93801	
		0.41488			0.43166
P 19	406.47798		P 26	403.50635	
		0.41732			0.43402
P 20	406.06066		P 27	403.07233	

First Order Wave Numbers of the Rotational Lines of the $2\nu_1$ Band of $N_2^{14}O^{16}$

$2\nu_1$ band of N_2O occurring in the sixth order

Line J	First Order Wave Number (vac cm⁻¹)	$\Delta\nu$	Line J	First Order Wave Number (vac cm⁻¹)	$\Delta\nu$
R 56	433.31248		R 36	431.59172	
		0.07563			0.09767
R 55	433.23685		R 35	431.49405	
		0.07672			0.09880
R 54	433.16013		R 34	431.39525	
		0.07781			0.09993
R 53	433.08232		R 33	431.29532	
		0.07889			0.10104
R 52	433.00343		R 32	431.19428	
		0.07998			0.10218
R 51	432.92345		R 31	431.09210	
		0.08107			0.10328
R 50	432.84238		R 30	430.98882	
		0.08215			0.10444
R 49	432.76023		R 29	430.88438	
		0.08325			0.10555
R 48	432.67698		R 28	430.77883	
		0.08436			0.10670
R 47	432.59262		R 27	430.67213	
		0.08545			0.10781
R 46	432.50717		R 26	430.56432	
		0.08655			0.10895
R 45	432.42062		R 25	430.45537	
		0.08765			0.11009
R 44	432.33297		R 24	430.34528	
		0.08877			0.11121
R 43	432.24420		R 23	430.23407	
		0.08988			0.11235
R 42	432.15432		R 22	430.12172	
		0.09097			0.11349
R 41	432.06335		R 21	430.00823	
		0.09210			0.11463
R 40	431.97125		R 20	429.89360	
		0.09322			0.11577
R 39	431.87803		R 19	429.77783	
		0.09431			0.11688
R 38	431.78372		R 18	429.66095	
		0.09544			0.11805
R 37	431.68828		R 17	429.54290	
		0.09656			0.11917
R 36	431.59172		R 16	429.42373	

$2\nu_1$ band of N_2O occurring in the sixth order

Line J	First Order Wave Number (vac cm^{-1})	$\Delta\nu$	Line J	First Order Wave Number (vac cm^{-1})	$\Delta\nu$
R 16	429.42373		P 4	426.65798	
		0.12031			0.14421
R 15	429.30342		P 5	426.51377	
		0.12145			0.14534
R 14	429.18197		P 6	426.36843	
		0.12260			0.14646
R 13	429.05937		P 7	426.22197	
		0.12372			0.14760
R 12	428.93565		P 8	426.07437	
		0.12490			0.14872
R 11	428.81075		P 9	425.92565	
		0.12598			0.14987
R 10	428.68477		P 10	425.77578	
		0.12715			0.15098
R 9	428.55762		P 11	425.62480	
		0.12830			0.15210
R 8	428.42932		P 12	425.47270	
		0.12944			0.15323
R 7	428.29988		P 13	425.31947	
		0.13056			0.15435
R 6	428.16932		P 14	425.16512	
		0.13170			0.15547
R 5	428.03762		P 15	425.00965	
		0.13285			0.15658
R 4	427.90477		P 16	424.85307	
		0.13399			0.15772
R 3	427.77078		P 17	424.69535	
		0.13511			0.15882
R 2	427.63567		P 18	424.53653	
		0.13627			0.15995
R 1	427.49940		P 19	424.37658	
		0.13740			0.16105
R 0	427.36200		P 20	424.21553	
		0.27820			0.16215
P 1	427.08380		P 21	424.05338	
		0.14082			0.16328
P 2	426.94298		P 22	423.89010	
		0.14193			0.16437
P 3	426.80105		P 23	423.72573	
		0.14307			0.16548
P 4	426.65798		P 24	423.56025	

$2\nu_1$ band of N_2O occurring in the sixth order

Line J	First Order Wave Number (vac cm^{-1})	$\Delta\nu$	Line J	First Order Wave Number (vac cm^{-1})	$\Delta\nu$
P 24	423.56025		P 40	420.76417	
		0.16658			0.18390
P 25	423.39367		P 41	420.58027	
		0.16769			0.18495
P 26	423.22598		P 42	420.39532	
		0.16878			0.18602
P 27	423.05720		P 43	420.20930	
		0.16988			0.18705
P 28	422.88732		P 44	420.02225	
		0.17097			0.18812
P 29	422.71635		P 45	419.83413	
		0.17207			0.18916
P 30	422.54428		P 46	419.64497	
		0.17315			0.19020
P 31	422.37113		P 47	419.45477	
		0.17425			0.19124
P 32	422.19688		P 48	419.26353	
		0.17531			0.19226
P 33	422.02157		P 49	419.07127	
		0.17642			0.19332
P 34	421.84515		P 50	418.87795	
		0.17748			0.19433
P 35	421.66767		P 51	418.68362	
		0.17855			0.19535
P 36	421.48912		P 52	418.48827	
		0.17964			0.19639
P 37	421.30948		P 53	418.29188	
		0.18070			0.19740
P 38	421.12878		P 54	418.09448	
		0.18178			0.19840
P 39	420.94700		P 55	417.89608	
		0.18283			0.19941
P 40	420.76417		P 56	417.69667	
					0.20044
			P 57	417.49623	

APPENDIX III. 25

First Order Wave Numbers of the Rotational Lines of the $(\nu_1 + 2\nu_2)$ Band of $N_2^{14}O^{16}$

$\nu_1 + 2\nu_2$ band of N_2O occurring in the sixth order

Line J	First Order Wave Number (vac cm^{-1})	$\Delta\nu$	Line J	First Order Wave Number (vac cm^{-1})	$\Delta\nu$
R 55	417.57487		R 35	415.14527	
		0.11610			0.12695
R 54	417.45877		R 34	415.01832	
		0.11672			0.12740
R 53	417.34205		R 33	414.89092	
		0.11730			0.12785
R 52	417.22475		R 32	414.76307	
		0.11788			0.12829
R 51	417.10687		R 31	414.63478	
		0.11847			0.12873
R 50	416.98840		R 30	414.50605	
		0.11907			0.12917
R 49	416.86933		R 29	414.37688	
		0.11961			0.12958
R 48	416.74972		R 28	414.24730	
		0.12020			0.13002
R 47	416.62952		R 27	414.11728	
		0.12077			0.13041
R 46	416.50875		R 26	413.98687	
		0.12132			0.13082
R 45	416.38743		R 25	413.85605	
		0.12186			0.13122
R 44	416.26557		R 24	413.72483	
		0.12242			0.13160
R 43	416.14315		R 23	413.59323	
		0.12293			0.13198
R 42	416.02022		R 22	413.46125	
		0.12349			0.13235
R 41	415.89673		R 21	413.32890	
		0.12398			0.13272
R 40	415.77275		R 20	413.19618	
		0.12452			0.13308
R 39	415.64823		R 19	413.06310	
		0.12500			0.13343
R 38	415.52323		R 18	412.92967	
		0.12551			0.13379
R 37	415.39772		R 17	412.79588	
		0.12599			0.13413
R 36	415.27173		R 16	412.66175	
		0.12646			0.13447
R 35	415.14527		R 15	412.52728	

$\nu_1 + 2\nu_2$ band of N_2O occurring in the sixth order

Line J	First Order Wave Numbers (vac cm^{-1})	$\Delta\nu$	Line J	First Order Wave Numbers (vac cm^{-1})	$\Delta\nu$
R 15	412.52728		P 5	409.63180	
		0.13480			0.14112
R 14	412.39248		P 6	409.49068	
		0.13513			0.14138
R 13	412.25735		P 7	409.34930	
		0.13547			0.14170
R 12	412.12188		P 8	409.20760	
		0.13576			0.14197
R 11	411.98612		P 9	409.06563	
		0.13610			0.14226
R 10	411.85002		P 10	408.92337	
		0.13640			0.14255
R 9	411.71362		P 11	408.78082	
		0.13672			0.14285
R 8	411.57690		P 12	408.63797	
		0.13702			0.14315
R 7	411.43988		P 13	408.49482	
		0.13731			0.14345
R 6	411.30257		P 14	408.35137	
		0.13764			0.14375
R 5	411.16493		P 15	408.20762	
		0.13791			0.14405
R 4	411.02702		P 16	408.06357	
		0.13822			0.14435
R 3	410.88880		P 17	407.91922	
		0.13850			0.14467
R 2	410.75030		P 18	407.77455	
		0.13882			0.14498
R 1	410.61148		P 19	407.62957	
		0.13908			0.14532
R 0	410.47240		P 20	407.48425	
		0.27905			0.14562
P 1	410.19335		P 21	407.33863	
		0.13997			0.14595
P 2	410.05338		P 22	407.19268	
		0.14023			0.14628
P 3	409.91315		P 23	407.04640	
		0.14053			0.14662
P 4	409.77262		P 24	406.89978	
		0.14082			0.14695
P 5	409.63180		P 25	406.75283	

$\nu_1 + 2\nu_2$ band of N_2O occurring in the sixth order

Line J	First Order Wave Number (vac cm^{-1})	$\Delta\nu$	Line J	First Order Wave Number (vac cm^{-1})	$\Delta\nu$
P 25	406.75283		P 41	404.35047	
		0.14730			0.15365
P 26	406.60553		P 42	404.19682	
		0.14765			0.15412
P 27	406.45788		P 43	404.04270	
		0.14801			0.15457
P 28	406.30987		P 44	403.88813	
		0.14837			0.15503
P 29	406.16150		P 45	403.73310	
		0.14875			0.15552
P 30	406.01275		P 46	403.57758	
		0.14912			0.15600
P 31	405.86363		P 47	403.42158	
		0.14950			0.15646
P 32	405.71413		P 48	403.26512	
		0.14988			0.15697
P 33	405.56425		P 49	403.10815	
		0.15030			0.15747
P 34	405.41395		P 50	402.95068	
		0.15068			0.15795
P 35	405.26327		P 51	402.79273	
		0.15109			0.15846
P 36	405.11218		P 52	402.63427	
		0.15148			0.15899
P 37	404.96070		P 53	402.47528	
		0.15192			0.15948
P 38	404.80878		P 54	402.31580	
		0.15233			0.15998
P 39	404.65645		P 55	402.15582	
		0.15277			0.16052
P 40	404.50368		P 56	401.99530	
		0.15321			0.16105
P 41	404.35047		P 57	401.83425	